T0310432

DRIVING CONTINUOUS PROCESS SAFETY IMPROVEMENT FROM INVESTIGATED INCIDENTS

Relation of this Book to Other Publications of the Center for Chemical Process Safety

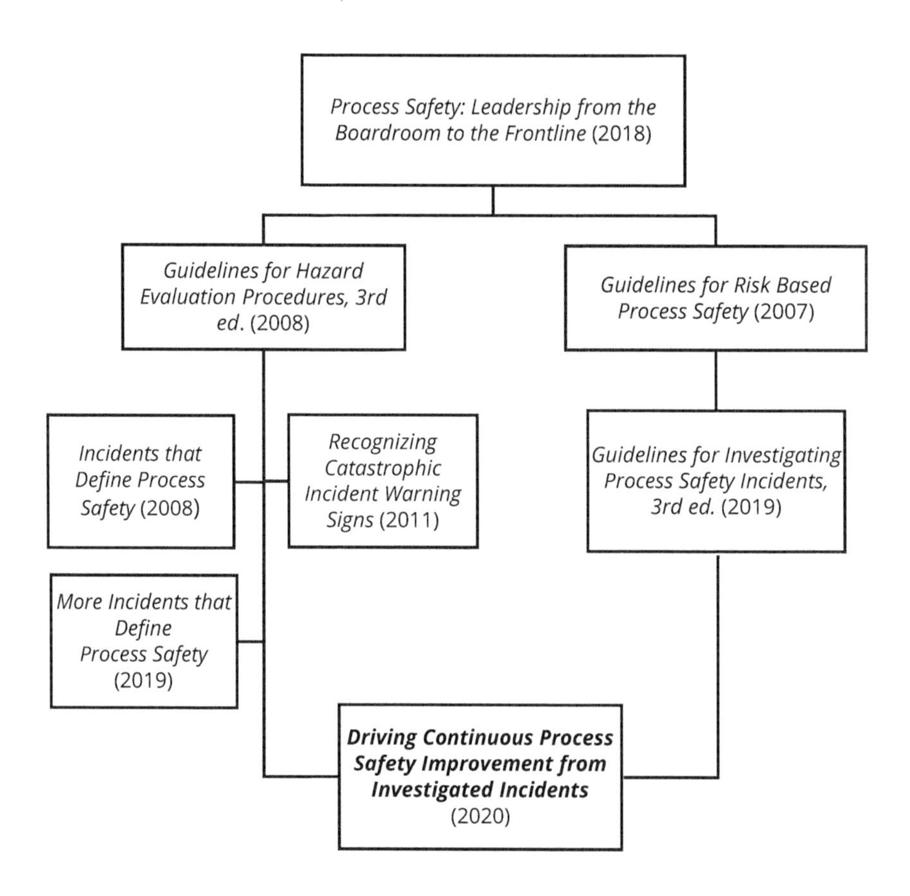

For a complete listing of CCPS books, please visit www.wiley.com/go/ccps.

DRIVING CONTINUOUS PROCESS SAFETY IMPROVEMENT FROM INVESTIGATED INCIDENTS

CENTER FOR CHEMICAL PROCESS SAFETY
of the
AMERICAN INSTITUTE OF CHEMICAL ENGINEERS
New York, NY

This edition first published 2021

A Joint Publication of the American Institute of Chemical Engineers and John Wiley & Sons, Inc.

Registered Office
John Wiley & Sons, Inc., 111 River Street, Hoboken, NJ 07030, USA

Editorial Office
111 River Street, Hoboken, NJ 07030, USA

For details of our global editorial offices, customer services, and more information about Wiley products visit us at www.wiley.com.

Wiley also publishes its books in a variety of electronic formats and by print-on-demand. Some content that appears in standard print versions of this book may not be available in other formats.

Library of Congress Cataloging-in-Publication Data

Names: American Institute of Chemical Engineers. Center for Chemical Process Safety, author.
Title: Driving continuous process safety improvement from investigated incidents / CCPS, American Institute of Chemical Engineers.
Description: First edition. | Hoboken, NJ : Wiley, [2021] | Includes bibliographical references and index.
Identifiers: LCCN 2020047579 (print) | LCCN 2020047580 (ebook) | ISBN 9781119768661 (hardback) | ISBN 9781119768678 (adobe pdf) | ISBN 9781119768685 (epub)
Subjects: LCSH: Chemical processes–Safety measures. | Chemical plants–Accidents.
Classification: LCC TP150.S24 D75 2020 (print) | LCC TP150.S24 (ebook) | DDC 660–dc23
LC record available at https://lccn.loc.gov/2020047579
LC ebook record available at https://lccn.loc.gov/2020047580

Cover Design: Wiley
Cover Image: © Trinity Mirror/Alamy Stock Photo

SKY10025736_032321

Disclaimer

ABOUT AIChE AND CCPS

The American Institute of Chemical Engineers (AIChE) has led efforts to improve process safety in the chemical, petroleum, and allied industries for more than six decades. Through strong ties with process designers, constructors, operators, maintenance professionals, safety professionals, and members of academia, AIChE has enhanced communications and fostered continuous improvement of the industry's high process safety standards. AIChE publications and symposia have become the premier information resources for those devoted to process safety and environmental protection.

AIChE formed the Center for Chemical Process Safety (CCPS) in 1985 after the tragic incidents in late 1984 in Mexico City, Mexico, and Bhopal, India. CCPS is chartered to develop and disseminate technical and leadership guidance and expertise to help prevent fires, explosions, toxic releases, and major environmental impacts.

CCPS is supported by more than 200 member-companies around the world. Members provide the necessary funding and professional expertise to many CCPS committees. A major CCPS product has been a series of Guidelines and Concept Books to assist those implementing the elements of process safety and risk management systems. This book is part of that series.

Dedication

CCPS and the members of the Learning from Investigated Incidents subcommittee dedicate this book to Mr. Adrian L. Sepeda, PE. Adrian has served CCPS in many capacities over the years. A mechanical engineer by training, he served as the Technical Steering Committee representative for Diamond Shamrock and then OxyChem; an original co-chair and then leader and editor for the *Process Safety Beacon* newsletter, helping it reach a monthly circulation of more than a quarter of a million worldwide; a co-developer of CCPS's Fundamentals of Process Safety course, which has given more than 10,000 industry professionals a firm grounding in process safety; and the driving force behind the creation and expansion of the CCPS Process Safety Incident Database (PSID), which now contains more than 850 investigated incidents that members can use to help drive continuous process safety improvement, as discussed in this book. He also served as a staff consultant on many CCPS book projects. His tireless efforts and selfless dedication have made CCPS a better, stronger, and more influential organization.

Louisa A. Nara, CCPSC
Global Technical Director, CCPS

TABLE OF CONTENTS

ACRONYMS AND ABBREVIATIONS

ACC	American Chemistry Council
ADKAR®	Awareness, Desire, Knowledge, Ability, Reinforcement (Learning Model)
AFPM	Association of Fuels and Petrochemicals Manufacturers
AIChE	American Institute of Chemical Engineers
ANP	Agência Nacional do Petróleo
API	American Petroleum Institute (USA)
ARIA	Analysis, Research, and Information on Accidents (Database)
ASME	American Society of Mechanical Engineers
BLEVE	Boiling Liquid Expanding Vapor Explosion
CCPS	Center for Chemical Process Safety
CSB	US Chemical Safety and Hazard Investigation Board
CUI	Corrosion Under Insulation
DIERS	Design Institute for Emergency Relief Systems
DSB	Dutch Safety Board
EASHW	European Agency for Health and Safety at Work
eMARS	European Commission Major Accident Reporting System
EPA	Environmental Protection Agency (USA)
EPICEA	Études de Prévention par l'Informatisation des Comptes Rendus d'Accidents (Database)
EPSC	European Process Safety Centre
FAA	US Department of Transportation, Federal Aviation Administration
HIRA	Hazard Identification and Risk Analysis
HR	Human Resources
HSE	Health, Safety, and Environment(al), or the Health and Safety Executive (UK) [usage depends on context]

ICC	Indian Chemical Council
IChemE	Institution of Chemical Engineers (UK)
IHLS	Independent High-Level Switch
IOGP	International Association of Oil & Gas Producers
ITPM	Inspection, Testing, and Preventive Maintenance
LOPC	Loss of Primary Containment
MIC	Methyl Isocyanate
MOC	Management of Change
NASA	National Aeronautical and Space Administration
NDOJ	National Diet of Japan
NFPA	National Fire Protection Association
NISA	Nuclear & Industrial Safety Agency (Japan)
OSHA	Occupational Safety and Health Administration (USA)
PDCA	Plan-Do-Check-Act
PHA	Process Hazard Analysis
PPE	Personal Protective Equipment
PSID	CCPS Process Safety Incident Database
PSMS	Process Safety Management System
PSSR	Pre-startup Safety Review
RBPS	Risk Based Process Safety
REAL	Recalling Experiences and Applied Learning (Model)
SADT	Self-Accelerating Decomposition Temperature
SHE	Safety, Health, and Environment(al)
SIS	Safety Instrumented System
TMR	Time to Maximum Rate
UK	United Kingdom
UKDOE	UK Department of Employment
US(A)	United States (of America)
USD	US Dollars

ACKNOWLEDGEMENTS

The American Institute of Chemical Engineers (AIChE) and its Center for Chemical Process Safety (CCPS) thank the Learning from Investigated Incidents Subcommittee members and their CCPS member companies for their generous support and technical contributions to this book. CCPS also thanks the members of the CCPS Technical Steering Committee for their advice and support.

CCPS Learning from Investigated Incidents Subcommittee

The Chair of the Learning from Investigated Incidents Subcommittee was Fred Henselwood of NOVA Chemicals. The CCPS staff consultant was Dan Sliva. Dr. Anil Gokhale, PE stepped into that role in the last few months of the project. CCPS thanks the entire subcommittee whose members were:

Fred Henselwood	NOVA Chemicals, Chair
Elena Blardony-Arranz	Repsol
Don Connolley	AcuTech Consulting Group
Rajender Dahiya	AIG
Alexandre Glitz	CCPS Emeritus
Allison Knight	3M
Jeffrey Lu	Wanhua Chemical Group Co., Ltd
Reid McPhail	Canadian Natural Resources Ltd
Gene Meyer	Kraton
Dan Miller	CCPS Emeritus
Pamela Nelson	ioMosaic
Janet Persechino	Enbridge
Sufyan Nor	Petronas
Gill Sigmon	AdvanSix
Ken Tague	CCPS Emeritus
Elliot Wolf	Chemours

CCPS thanks Scott Berger and Associates, LLC for preparing the manuscript. Scott Berger, President, served as project manager and co-writer. Kristine Chin served as co-writer. Cynthia Berger, member of the National Association of Science Writers, served as editor.

CCPS also is grateful to John Herber, CCPS Emeritus, for his help evaluating publicly reported incidents for the index.

Peer Reviewers

Before publication, all CCPS books receive a thorough peer review. CCPS gratefully acknowledges the thoughtful comments and suggestions of the peer reviewers. Their work enhanced the accuracy and clarity of content in this publication.

Paul Amyotte	Dalhousie University
David Donnick	AdvanSix
Emmanuelle Hagey	NOVA Chemicals
Swapan Kumar Hazra	Former Chair, ICC (India) SHE Committee
David Heller	AcuTech Consulting Group
John Herber	CCPS Emeritus
Katherine Prem	LyondellBasell
Deepak Sharma	Bayer
Bruce Vaughen	CCPS

GLOSSARY

Readers may be accustomed to using the term "lessons-learned" to describe what they have learned from an incident. This book, however, follows the standard CCPS Glossary spelling and definition:

Lessons Learned: Applying knowledge gained from past incidents in current practices.

In other words, learning has occurred only when the company or plant has applied and institutionalized the knowledge gained. Applying knowledge means making changes to the company's and/or plant's PSMS, standards, and policies and implementing these changes. Institutionalizing knowledge means assuring the changes remain in place.

The term *legacy* is use in this book to refer to equipment designed to an earlier version of a standard. Some standards will contain a *legacy clause* deeming such equipment to comply with the standard, even if their design would not comply if designed that way today. Such equipment is considered *legacied*. These terms are used in place of the now-obsolete terms *grandfather*, *grandfather clause*, and *grandfathered*.

The rest of these terms follow the CCPS glossary, which may be accessed online:

www.aiche.org/ccps/resources/glossary

FOREWORD

"We now accept the fact that learning is a lifelong process of keeping abreast of change. And the most pressing task is to teach people how to learn."—Peter Drucker, Management Consultant and Educator

This quote concisely captures the essence of the book you are about to read. As an engineering educator and process safety researcher, I have experienced several watershed moments that have given me clarity of thought and shaped the direction of my professional life. Two experiences that stand out were reading the treatment by Professor Richard Felder of North Carolina State University on learning styles and learning objectives and the writings of renowned safety expert Trevor Kletz on the practicalities of inherently safer design.

I now have an additional paradigm-shifting perspective for which I thank the authors of this book and the CCPS subcommittee members and staff. They have done a marvelous job of identifying a knowledge gap and providing the means to close it. The subcommittee has married the process of learning to the core principles of process safety in the Recalling Experiences and Applied Learning (REAL) Model, described in this book. Implementing the guidance and the REAL Model in your own facility and company will bring significant benefits to you, your co-workers, and the communities in which you work and live.

The introductory chapter describes the motivation for studying case histories of process incidents: these events continue to occur despite our collective efforts to improve technology, standards, and management systems. Other books focus on descriptions of the incidents and their intrinsic learning value—such as the lessons learned from external incidents. This book is different. It focuses on the learning process and how lessons from external incidents—as well as from internal incidents and near-misses—can be embedded in the memory of an organization.

By its very nature, the study of incident investigation reports is a study in failure. Root cause analysis, if done properly, will identify deficiencies in the application of process safety management system elements and the adoption of core process safety concepts. But this is only half the story. We need to know how things can go right, not just how they can go wrong. It is imperative to know the failures that occurred at Flixborough, Bhopal, and Texas City, but it is equally imperative to know the successes illustrated by the six scenarios described in Chapters 9-14. Although these scenarios are fictional, they are based on a range of real-world situations you might encounter in your own facility, and they lay out effective real-world strategies for addressing all-too-common challenges – and ensuring that problems don't recur.

There is a strong and inescapable human component to process safety. Learning from process incidents—whether internal or external to a company—is the responsibility of everyone from design and operations personnel to managers and corporate board members. *Driving Continuous Process Safety Improvement from Investigated Incidents* gives each of us a much-needed roadmap for success in transforming lessons learned into institutional knowledge. Simply put, this book represents a step change in our approach to preventing process incidents and ensuring safer workplaces.

Paul Amyotte, PEng
Professor of Chemical Engineering & C. D. Howe Chair in Process Safety
Dalhousie University, Halifax, NS, Canada
July 2020

EXECUTIVE SUMMARY

"Those who don't know history are doomed to repeat it."
—Edmund Burke, Anglo-Irish Statesman

In February 1985, two months after the Bhopal tragedy, senior executives from prominent chemical, petroleum, pharmaceutical, and engineering companies asked the American Institute of Chemical Engineers (AIChE) to launch a global cross-industry collaborative effort to improve process safety. With their companies' support, the Center for Chemical Process Safety (CCPS) was formed in March 1985.

Since then, CCPS has both benefitted from and helped increase the collective knowledge of its members and the whole industry, producing more than 100 publications that are used by companies around the world to improve their process safety efforts. Continuously seeking to improve process safety performance is a hallmark of CCPS's Risk Based Process Safety (RBPS) approach. In fact, Management Review and Continuous Improvement is an explicit element of RBPS.

Other organizations have also contributed to improvements in process safety. Prominent among them are several national, government-sponsored investigative organizations. These organizations investigate incidents that have high learning potential, then communicate their findings to relevant regulatory authorities and to the whole industry.

The reports and videos these organizations produce are a treasure trove to companies seeking to improve their process safety performance. These products can be the source of corporate process safety improvements, reducing the risk of a tragic loss or even a sobering near-miss—and they are available for free.

This book presents a novel approach to learning from publicly investigated incidents. The Recalling Experiences and Applied Learning (REAL) Model builds on the traditional method for learning from internal incident investigations. Readers can, and should, also use the REAL Model to enhance their learning from internal incidents and to strengthen their recommendations and ongoing communication efforts. The eight steps of the REAL Model—and where they fit in the traditional Plan-Do-Check-Act improvement cycle—are summarized in Figure FM.1.

	Individuals	Company
Gather facts	2. Seek learnings 3. Understand 4. Drilldown **Do**	1. Focus **Plan**
Interpret and act	5. Internalize 6. Prepare **Check**	8. Embed and refresh 7. Implement **Act**

Figure FM.1 Recalling Experiences and Applied Learning (REAL) Model

The book starts by highlighting the importance of driving permanent change based on learning from incidents. It then lays the foundation for the REAL Model, which is then introduced, followed by a discussion of learning styles and how to leverage them to effectively communicate what has been learned and keep the learning fresh. The latter part of the book discusses landmark incidents and features 6 hypothetical scenarios that are based on real-world situations readers may encounter. The book concludes with a call-to-action to drive continuous improvement. A brief description of each chapter follows.

- *Chapter 1* explains why it is important to translate the findings from incidents into lessons learned that become a part of the corporate culture.
- *Chapter 2* summarizes learning opportunities that are often overlooked and lists valuable resources including databases, publications, and more.
- *Chapter 3* evaluates obstacles to learning and describes a general philosophy for overcoming those obstacles.
- *Chapter 4* provides examples of incidents where companies failed to learn from previous incidents, whether in their own companies or externally.
- *Chapter 5* examines literature about learning, considering both how individuals and companies learn and change. Based on this literature, the

Recalling Experiences and Applied Learning (REAL) Model for learning from incidents is introduced.

- *Chapter 6* describes the REAL Model in detail, providing a roadmap for learning from incidents.
- *Chapter 7* discusses how to use the full range of learning styles in your efforts to keep lessons learned fresh and to prevent erosion of knowledge and normalization of deviance.
- *Chapter 8* describes landmark incidents that contributed significantly to our understanding of process safety. In your efforts to drive process safety improvement, it is important to ensure that the lessons learned from these incidents have been well implemented and maintained throughout your organization.
- *Chapters 9–14* present a set of fictional but realistic scenarios that show how you might apply the REAL Model in your own workplace to translate the findings from incidents to corporate lessons learned.
- *Chapter 15* summarizes the concepts described in the book and challenges the reader to drive continuous improvement.

One noteworthy feature of this book is the Index of Publicly Evaluated Incidents, presented in the Appendix. Use this index to identify specific incidents with relevant findings that can help you advance your corporate improvement goals. Incidents described in the book that are included in the index are flagged with a text box, as shown at right. You can consult the Index of Publicly Evaluated Incidents to find a link to the incident report, along with an assessment (by a book committee member) of the most important findings.

See Appendix index entry XYZ

Thank you for reading this book. There is really no business priority that comes ahead of protecting workers, facilities, communities, and the environment. Process safety depends on all of us performing our roles with professionalism and competence, using the best knowledge available. Just as excellence in process safety cannot be achieved by one person alone, process safety knowledge cannot be held only by company experts. When it comes to process safety knowledge, the only thing that matters is what we know, and do, and manage collectively.

APPLICABILITY OF THIS BOOK

This book was written for anyone working in the process industries to prevent major incidents—incidents that could potentially injure or cause the death of many people, destroy company assets and nearby property, and harm the environment. The range of industries that may benefit from this book extends well beyond the traditional home of process safety—chemicals, oil and gas production and refining, and pharmaceuticals—to include:

- ammonia refrigeration
- food, beverage, and plant-based material processing
- industries that produce combustible dust, including metal smelting and re-melting and fabrication of resin-impregnated fibers
- materials processing, electronics, and chemical vapor deposition
- mining, paper, and plastic and resin molding
- terminals and storage and distribution facilities
- water and wastewater treatment.

Picture ethylene glycol leaking from a cooling jacket into a batch of beer, a hexane spill from soybean oil extraction that catches fire, a pump seal leak in a terminal that sets an entire tank farm on fire, or a layer of iron powder or fiberglass resin dust deposited on work surfaces that becomes suspended and explodes. While they didn't happen in the chemical or petroleum industries, these are all process safety incidents that everyone can learn from, just as the companies in these industries can learn from chemicals and petroleum.

Going beyond industrial processes, an airplane crash, a dam failure, a spacecraft that burns up, a nuclear plant that melts down, and yes, even the novel coronavirus spreading across the world as this book was being written are disasters that were preceded by events we could have learned from—but didn't. All of them present valuable opportunities for anyone who manages process safety hazards to learn and drive improvement.

Set aside the notion that what happened in a different sector does not apply to your company. Whatever your industry sector, you can learn from other sectors. Flammable liquids and gases don't care what industry they're used in, they are just as flammable everywhere. Reactive or unstable chemicals don't runaway only in chemical plants. Combustible dust can explode even in locations where national regulations don't control this hazard. Regardless of sector, hazards must be controlled by barriers, and a well-functioning process safety management system must ensure the reliability of those barriers.

Electronic materials accompanying this book

The Index of Publicly Evaluated Incidents included in the Appendix is also available in MS Excel® spreadsheet form. The spreadsheet has enhanced lookup capabilities and may be updated by CCPS periodically to include newly investigated incidents. Download the spreadsheet from the CCPS website:

www.aiche.org/ccps/publications/learning-incidents

After downloading, open the spreadsheet and enter the password CCPSLearning to access it.

You are also invited to help CCPS add new external incidents to this database; the above web page includes an indexing form you can complete. Email the completed indexing form to CCPS@AIChE.org.

1
INTRODUCTION

"Learning is not compulsory... neither is survival."
—W. Edwards Deming, Engineer and Management Consultant

Nearly everything we do today, as we manage process safety to prevent losses of primary containment that result in fires, explosions, and toxic releases, we do because of conditions that led to past incidents. Our engineering forebears began building the modern practice of process safety at the beginning of the industrial revolution. Subsequent generations have steadily advanced process safety.

For example, when E.I. DuPont built a black powder works in Delaware, USA, in 1802, he took note of the explosions that had happened in other black powder works. To protect his workers, family, and property, his process buildings were constructed of thick stone, with blow-out walls aimed away from people and buildings (Klein 2009).

Similarly, Sir Humphrey Davy noted the large number of coal dust explosions in English mines in the early nineteenth century (Gibbs 2020). After talking to miners who survived such explosions, he designed an explosion-proof lamp based on principles still used today in flame arrestors and explosion-proof electrical boxes (Figure 1.1).

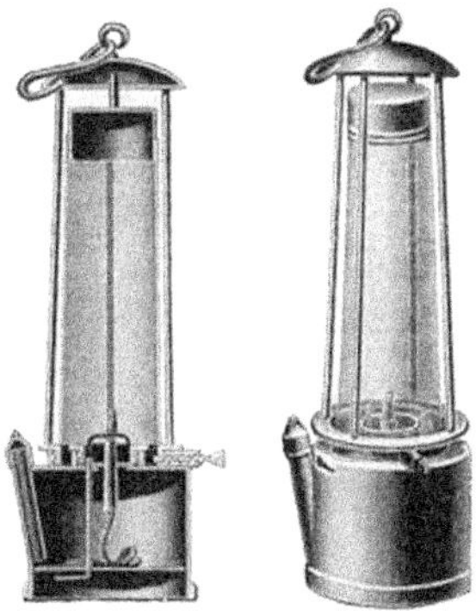

Figure 1.1 The Davy Lamp

In 1880, H.R. Worthington, A.L. Holley, and J.E. Sweet founded the American Society of Mechanical Engineers (ASME) to create uniform engineering standards that would ensure safety, reliability, and efficiency (ASME 2020).

Working on behalf of the chemical engineering profession, the American Institute of Chemical Engineers (AIChE) began to share findings and recommendations from process safety incidents via the Ammonia Plant Safety (Williams 2005) and Loss Prevention Symposia in the 1950s and 1960s (Freeman 2016). AIChE's Design Institute for Emergency Relief Systems (DIERS) began publishing guidelines for multiphase relief systems in the 1970s (AIChE 2020a).

Until the mid-1980s, institutional lessons learned came in the form of technology innovations, new or revised standards and codes, or back-up systems. This began to change with the formation of AIChE's Center for Chemical Process Safety (CCPS) in 1985. CCPS began the process of formally leveraging incident findings and successful practices into "Guidelines" and "Concepts" (Berger 2009). In 1988 CCPS codified the first Process Safety Management System (PSMS). The CCPS 12 Elements (CCPS 1989) provided the first organized common framework to comprehensively manage all the standards, technologies, and practices needed to control a company's process safety hazards. The original framework has evolved into today's 20 elements of Risk Based Process Safety (RBPS), which are organized in four pillars: Commit to Process Safety, Understand Hazards and Risk, Manage Risk, and Learn from Experience (CCPS 2007).

Regulations around the world also began to emerge in the 1980s, most notably the Sevesso Directive in the European Union, the Process Safety Management (PSM) regulation in the USA, and the Control of Major Accident Hazards (COMAH) in the UK. Most national and regional process safety regulations are based on one or a combination of these original regulations.

Unfortunately, incidents continue to happen despite 200 years of continuous development of technology, standards, publications, and management systems. They continue to happen despite the great number of recommendations from incident investigations conducted by every operating company in this industry. And nearly every incident that occurs in an industry, a company, or a plant has root causes that resemble the causes of previous incidents.

1.1 The Focus of this Book

CCPS (CCPS 2019a) and others have written guidelines addressing the general process of incident investigations. These books focus heavily on the process of investigation, the determination of root causes and causal factors, and the process of developing findings and recommendations. CCPS and others also have published books that describe past incidents to extract the lessons that could be learned from them (Gil 2008; CCPS 2019b; Kletz 2019; Hopkins 2008; Hopkins 2012). What's more, CCPS provides several publications addressing how to drive a culture of improvement in process safety (CCPS 2018; 2019c).

Just the same, incidents that look the same as previous incidents continue to occur—whether they happen at a site, or within a company, or replicate well publicized external incidents. Section 3.2 will discuss the numerous reasons companies fail to learn, including but not limited to:

- imbalance between production and safety
- corporate culture problems
- employment turnover
- financial or liability concerns
- lack of employee involvement
- lack of leadership ownership of process safety
- lack of sense of vulnerability
- knowledge remaining in silos
- normalization of deviance.

This book seeks to help companies overcome the reasons they fail to learn. It greatly expands on the process for:

- Seeking and obtaining key findings from external incidents.
- Translating findings into lessons learned. And especially
- Converting these lessons learned into institutional knowledge.

While the examples in this book focus on learning from incidents outside the company, the process described in this book can—and should—be applied to transform findings from internal incidents and near-misses into institutional knowledge.

Let's define some key terms for this book. Note that most of these terms can be found in the CCPS glossary, while a handful are specific to this book.

Causal factor: A major unplanned, unintended contributor to an incident (a negative event or undesirable condition) that, if eliminated, would have either prevented the incident or reduced its severity or frequency.

Root cause(s): A fundamental, underlying, system-related reason why an incident occurred that identifies a correctable failure(s) in management systems. There is typically more than one root cause for every process safety incident.

Findings: The root causes and causal factors of the incident, as determined by the investigator.

Near-miss: An event in which an accident (that is, property damage, environmental impact, or human loss) or an operational interruption could have plausibly resulted if circumstances had been slightly different.

Lessons learned: The process of applying knowledge gained from past incidents in current practices.

Institutional knowledge: The translation of the lessons learned by the experts into the company's standards and policies, its PSMS, its culture, and into the rigor and professionalism with which it manages process safety.

The difference between the latter two terms is critical in this book. Specifically, individuals do the primary learning, and then work with others in the organization to translate what they learned into improvements to 1) the corporate PSMS, 2) relevant standards and policies, and 3) designs and practices in company facilities. Only once these improvements are in place, are being followed, and are being managed on a reliable ongoing basis do lessons learned become institutional knowledge.

After making the case for continuous learning from incidents, we will evaluate ways that individuals and companies learn. We will then describe a recommended model for continuous corporate change driven from incidents, the Recalling Experiences and Applied Learning (REAL) Model. Next, we will provide a variety of scenarios that show companies in different industries applying this model to transform findings from incidents into institutional knowledge—and permanently retain that knowledge.

1.2 Why Should We Learn from Incidents?

In 1995 The Dow Chemical Company established a "Generational Goal" of reducing its process safety incidents by 90% between 1996 and 2005. Dow took many proactive steps including:

- implementing a rigorous process safety management system
- establishing centers of excellence
- defining key roles for leaders in driving process safety improvement
- learning from internal and external incidents.

Additional detail may be found in the reference (Champion 2017). Although Dow didn't quite meet its initial goal, the reduction in incidents was still quite significant. Inspired by the progress, Dow set a new goal: to reduce incidents by another 75% by 2015. However, by 2008 Dow realized that, for the past five years, their performance had effectively plateaued. Other members of the American Chemistry Council (ACC) had the same result, as shown by the dotted and dashed lines in Figure 1.2.

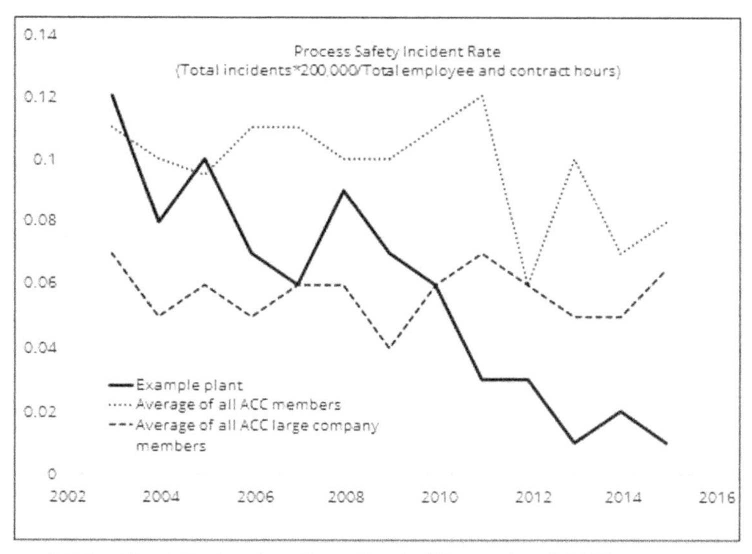

Figure 1.2 Incident Reduction Case Study (Champion 2017)

Dow analyzed the bottlenecks preventing further incident reduction and noted that many incidents had the same management system failures and causal factors: asset integrity and work practices related to corrosion under insulation (CUI). Dow sought out findings, recommendations, and best practices related to these factors from many internal and external sources. They translated what they learned into new standards, updated their PSMS, and drove training and knowledge about CUI across all their facilities. The solid line in Figure 1.2 shows the result at one plant, where the process safety incident rate dropped from 0.12 per 200,000 person-hours to 0.01 between 2003 and 2015.

With the current industry rate of process safety incidents, few days pass without a media report of a fire, explosion, or toxic release that has caused injuries or fatalities. We need to do better than we have been doing. And the

Dow Plant example shows that we can—by learning from incidents and embedding what we learn in the culture.

1.2.1 The Theory of Root Cause Correction

Most process safety hazards require multiple barriers to control them. No barrier is 100% reliable. Every barrier has a "probability of failure on demand," represented in Figure 1.3 as holes in slices of Swiss cheese.

- Hazards are controlled by multiple protective barriers.
- Barriers may have weaknesses or "holes."
- When the holes align, a hazard may pass through all the barriers, with the potential for adverse consequences.
- Barriers may be physical or engineered containment, or procedural controls dependent on people.
- Holes may be latent/incipient or opened by people.

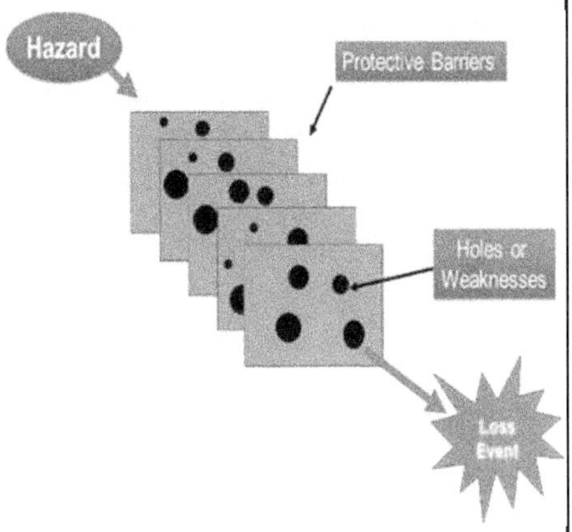

Figure 1.3 The Swiss Cheese Model (Adapted from Reason 1990)

Preventive barriers control process deviations and keep hazards within process equipment. When releases do occur, mitigating barriers prevent or reduce the consequences for workers, the facility, the public, and the environment.

If we do not manage our barriers properly, it's as if we increased the size of the holes in the Swiss cheese. The probability of failure on demand—the probability that hazards will pass through the barriers and cause harm—will increase.

We manage process barriers by adhering in a disciplined way to our PSMS and applicable standards. For example, the safe work practices element of RBPS (CCPS 2007) controls the following barriers:

- control of ignition sources
- hot work

- human factors
- isolation
- job hazard analysis/job safety analysis
- line and equipment opening
- lockout/tagout
- monitoring for flammables/toxic gases in work area
- non-routine work
- permits to work.

Very few process safety incidents occur for unforeseen reasons. In nearly every incident that has occurred, at least one, and usually more than one, important process barrier has failed. This means gaps existed in the PSMS and/or standards, or in the execution of the PSMS and/or standards.

Faithful and professional execution of the PSMS and adherence to standards should control and manage all barriers. When an incident has occurred, a proper incident investigation should identify the gaps in the PSMS and standards that allowed some barriers to fail. Closing these gaps will prevent the same incident from happening again. More importantly, closing these gaps makes progress toward eliminating all incidents that the existence of these gaps would allow.

1.2.2 Acting on Learning from High-Potential Near-misses

A near-miss is sometimes defined as an incident that, by chance, had no or only minimal consequences on people, the process, and the environment. As mentioned above, most process hazards require more than one barrier to control them. What if only one of these barriers fail, but the other(s) function? What if a confused operator reaches to open a valve and realizes just in time that it was the wrong valve? These, too, are near-misses. In all the following cases, a near-miss has occurred:

- any time your process has gone outside the safe operating window
- when testing or inspection reveals that a barrier has failed or has a latent defect
- when an incorrect action was stopped just in time.

> Near-misses are a gift — on a silver platter.

Just like actual incidents, near-misses happen because of gaps or weaknesses in PSMS, standards, and process design. Because near-misses point out these gaps without harming people or your process, they are a gift on a silver platter. Use this gift to find weaknesses in your company's PSMS, standards, culture, and knowledge. Develop actions broad and deep enough to correct these weaknesses. Ensure that the company implements these actions. Then ensure that weaknesses don't creep back in by using metrics and audits. Finally, keep everyone across the company aware of the consequences of tolerating gaps and weaknesses.

1.2.3 Learning from Other Company's (External) Incidents

A significant body of literature (and other media) describing the details of high-profile incidents has accumulated over the years. Many CCPS books and publications from other sources describe the conditions that led to incidents and the root causes, highlighting important lessons that companies should learn and embed.

The US Chemical Safety and Hazard Investigation Board (CSB) is one of the best-known sources of literature and videos about incidents. Many more sources exist, as Section 2.2 will cover in more detail.

By evaluating external incident reports, you will usually recognize root causes, findings, and recommendations that apply directly to your process, plant, or company. You can harvest them and begin to apply them immediately. As you institutionalize these lessons learned, you can use images from the reports to drive the message home visually.

The resources listed in Section 2.2 describe hundreds, perhaps thousands, of external incidents. Many of these incidents have been indexed in the Appendix to this book, along with their key root cause or causal factor as identified by the investigation team and the CCPS subcommittee members who championed this book. External incident reports are the gift that keeps on giving. If you haven't evaluated the learning opportunities from external incidents, or if you haven't evaluated them broadly or deeply enough, you have missed valuable opportunities to adapt and improve your PSMS and standards to prevent similar incidents from happening in your plant(s). There is no reason why your company should experience an incident with the same root cause as an external incident.

1.2.4 Societal Expectations and the Business Case

The CCPS guidelines for corporate process safety leadership detail many ways that society expresses its demand for industry to eliminate process safety incidents (CCPS 2019). Government agencies would not provide external incident reports for us to learn from if society didn't want us to learn and improve, nor would they enact and enforce process safety regulations. Similarly, community advocates press for new and stronger regulations, and victims seek civil penalties when incidents occur.

Society does not want companies to forego profitability, but instead wants profitability and protection go hand in hand. CCPS puts it this way:

> "A business case for process safety should not be necessary. The need to protect workers, the community, and the environment is as obvious as the need to protect market share, provide and protect shareholder value, and make payroll. But with the many daily demands of business, leaders can sometimes leave the obvious behind. When a company and its leaders leave process safety behind, they lose out on the significant financial and organizational benefits (CCPS 2019)."

CCPS further explains the business case for process safety, touching on five financial and organizational benefits:

- *corporate social responsibility*. Acting responsibly has a positive impact on share price.
- *loss prevention*. By avoiding incidents, the company avoids associated costs.
- *sustainable growth*. Process safety enhances the company's reliability and productivity.
- *business flexibility*. Easing societal resistance to the company's presence enables it to focus on growth.
- *leadership excellence*. Policies, practices, and competencies that drive rigor, dedication, and professionalism carry over to all other parts of the business.

In summary, society expects our industry to eliminate incidents and their impacts on workers, the public, and the environment. With strong leadership and continuous learning, the company can meet society's expectations while adding business value.

With all the external incidents available to learn from, we can meet both societal and shareholder expectations. We must study the findings of these incidents, build them into our PSMSs, standards, and policies, and make them part of our culture.

So, let's get to work.

1.3 References

1.1 AIChE (2020). *About DIERS*. www.aiche.org/diers/about (accessed May 2020).

1.2 ASME (2020). *History of ASME Standards*. asme.org/codes-standards/about-standards/history-of-asme-standards (accessed May 2020).

1.3 Berger, S.A. (2009). History of AIChE's Center for Chemical Process Safety. *Process Safety Progress* 28 (2): 124-127.

1.4 CCPS (1989). *Guidelines for Technical Management of Process Safety*. New York: AIChE.

1.5 CCPS (2007). *Guidelines for Risk Based Process Safety*. Hoboken, NJ: AIChE/Wiley.

1.6 CCPS (2018). *Essential Practices for Creating, Strengthening, and Sustaining Process Safety Culture*. Hoboken, NJ: AIChE/Wiley.

1.7 CCPS (2019a). *Guidelines for Investigating Process Safety Incidents*, 3rd Ed. Hoboken, NJ: AIChE/Wiley.

1.8 CCPS (2019b). *More Incidents that Define Process Safety*. Hoboken, NJ: AIChE/Wiley.

1.9 CCPS (2019c). *Process Safety Leadership from the Boardroom to the Frontline*. Hoboken, NJ: AIChE/Wiley.

1.10 Champion, J.W., Van Geffen, S., and Borrousch. L., (2017). Reducing process safety events: An approach proven by sustainable results. *Proceedings of the 13th Global Congress on Process Safety*, San Antonio, Texas (26–29 March 2017). New York: AIChE.

1.11 Freeman, R. (2016). History of the loss prevention symposium: The first 50 years—"there are no secrets in safety." *Process Safety Progress* 35 (1): inclusive pages 32–35.

1.12 Gibbs, F.W. (2019). Sir Humphrey Davy, in *Encyclopedia Britannica*, www.britannica.com/biography/Sir-Humphry-Davy-Baronet (accessed December 2019).

1.13 Gil, F. and Atherton, J. (2008). *Incidents that Define Process Safety*. Hoboken, NJ: CCPS/AIChE/Wiley.

1.14 Hopkins, A. (2008). *Failure to Learn: the BP Texas City Refinery Disaster*. New South Wales: CCH Australia Ltd.

1.15 Hopkins, A. (2012). *Disastrous Decisions: The Human and Organisational Causes of the Gulf of Mexico Blowout*. New South Wales: CCH Australia Ltd.

1.16 Klein, J.A. (2009). Two centuries of process safety at DuPont. *Process Safety Progress* 28 (2): 200–202.

1.17 Kletz, T. and Amyotte, P. (2019). *What Went Wrong*, 6th Ed. Oxford, UK: Butterworth-Heinemann.

1.18 Reason, J. (1990). The contribution of latent human failures to the breakdown of complex systems. *Philosophical Transactions of the Royal Society of London, Series B, Biological Sciences* 327 (1241): 475–484.

1.19 Williams, G.P. (2005). 50-year history of the AIChE. Ammonia Safety Symposium, *Proceedings of the Safety in Ammonia Plants and Related Facilities Conference*. Toronto: 25-29 September 2005.

2
LEARNING OPPORTUNITIES

"Study the past if you would define the future."
—Confucius, Chinese Philosopher

Continuous learning is key in the success of process safety, whether you're sitting in the corner office at corporate headquarters or in the control room at the plant. It is as true for companies as it is for individuals. This chapter will discuss learning opportunities, some of which will be familiar and others that you may have overlooked, as well as a wide range of resources you can use to increase your knowledge of process safety.

2.1 Think Broadly

When it comes to learning opportunities, think broadly. Don't confine yourself to a specific example and solution. Instead, look at the bigger picture. Look for multiple sources, scenarios, and solutions, and think beyond your own industry. The following sections contain several examples.

2.1.1 Looking Beyond the Specific Circumstances

Facility siting is a safety issue that continues to be overlooked in our industry. In the 2005 Texas City, TX, USA, incident, for example, trailers were destroyed during an explosion. Take this learning one step further, and consider that any building, not just trailers, can be impacted by an explosion, whether by the blast wave or by projectiles.

See Appendix index entry C11

This lesson could have been learned from the 1992 Castleford, UK, incident (HSE 1994). The energetic force of a jet fire from an open manway ripped through a

See Appendix index entry S7

control building, causing severe damage (Figure 2.1). The incident perfectly illustrates the concept of "avoiding the line of fire." Knowing about the control room in Castleford, planners could have situated the trailers at Texas City out of the line of fire.

Figure 2.1 Castleford Plant Control Room (Foreground) after Jet Fire (Source: HSE 1994)

See Appendix index entry C31

Here's another example of looking beyond specific circumstances. The 2016 Baton Rouge, LA, USA, incident (CSB 2017) that resulted in the ignition of an isobutane vapor cloud was caused by an employee confusing two different plug valve gear box designs being used in the same application. Although 97% of the gear boxes had modern designs that eliminated the potential for this kind of incident, the rest had a legacy design—one considered to meet the current standard because it had been designed to an older standard—that did not.

Knowing about this incident, you can broaden your learning. In general, you should ask yourself, "What equipment at my company might fit this scenario?" Look for situations in your facility where pieces of equipment in multiple designs serve the same purpose. Extend the concept even further to legacied equipment that may require special maintenance procedures.

2.1.2 Learning from Other Industries

Thinking beyond individual incidents, a broader perspective would be to look at similar incidents in a range of industries. Combustible dust explosions are a good example of a type of incident that can spark learning across industries. Figure 2.2 show the wide range of manufacturing sectors that have suffered combustible dust incidents (CSB 2006).

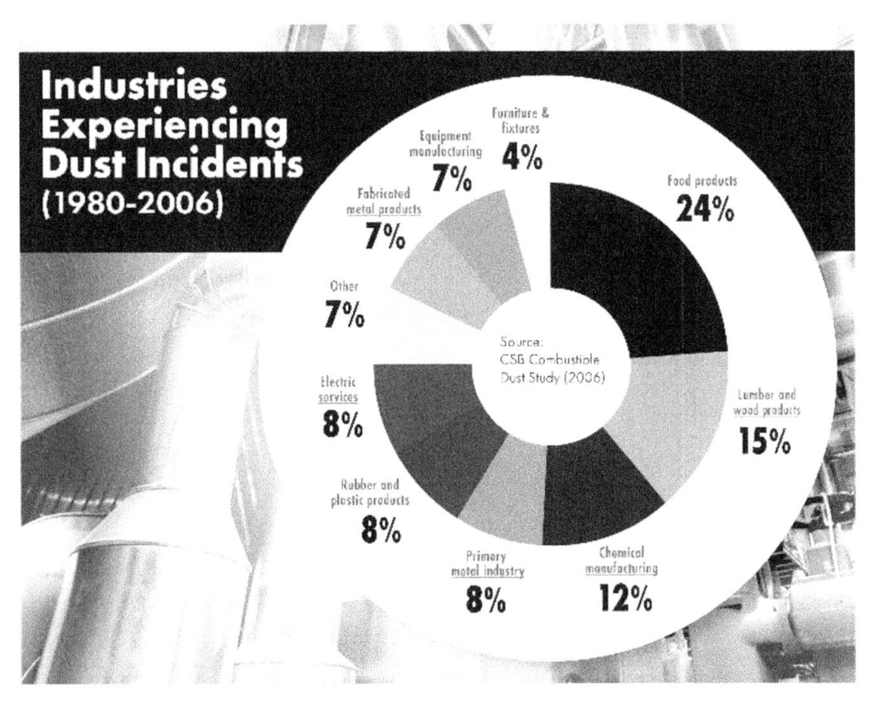

Figure 2.2 Combustible Dust Incidents Across Industries (Source CSB 2016)

Combustible dust explosions occur when fuel (e.g., sugar, flour, cornmeal, sawdust, chemical or pharmaceutical powder, and metal dust) is suspended in a confined area in the presence of an ignition source from hot work (e.g., spark or electrostatic) and oxygen. This can occur inside process and storage equipment or upon a release from equipment. Additionally, process dust that accumulates on floors, beams, and other horizontal surfaces may become suspended due to vibration or air flow. Greater confinement and congestion increase the severity of the explosion.

Recent examples of combustible dust explosions in various industries are shown in Table 2.1.

Table 2.1 Sampling of Dust Explosion Incidents Across Industries (Source: Adapted from CSB 2006)

Industry	Description	
Food	Date: 31 May 2017 Ignition source: Corn dust	CSB 2020a
	Combustible dust explosions from dry corn milling operation in Cambria, WI, USA, resulted in five fatalities and injured the remaining 14 workers. The investigation is ongoing.	
Metal recycling	Date: 9 December 2010 Ignition source: Titanium powder	See Appendix index entry C4
	An explosion ripped through a New Cumberland, WV, USA titanium plant, fatally injuring three workers.	
Rubber products	Date: 29 January 2003 Ignition source: Fine plastic powder	See Appendix index entry C75
	An explosion and fire destroyed a plant in Kinston, NC, USA causing six deaths and dozens of injuries. The facility produced rubber stoppers and other medical products. In a manufacturing area, fine plastic powder accumulated above a suspended ceiling and ignited.	
Auto parts	Date: 20 February 2003 Ignition source: Resin dust	See Appendix index entry C20
	An explosion and fire damaged a manufacturing plant in Corbin, KY, USA, fatally injuring seven. The facility produced fiberglass insulation for the automotive industry. The explosion was fueled by resin dust accumulated in a production area, likely ignited by flames from a malfunctioning oven.	

Your day-to-day work may not expose you to industries other than your own. Consider periodically scheduling time to refresh and exercise your mind. Each of CCPS's more than 100 books were, by design, prepared by committee members representing a range of industries and include as broad a range of learning opportunities as possible (CCPS 2020a). Reading any CCPS book is a good way to learn from industries and sectors different from your own. A useful starting point is *Guidelines for Risk Based Process Safety* (CCPS 2007), which offers thought-provoking ideas in every chapter.

2.1.3 Learning from Regulatory Standards and Beyond

Thinking broadly about the guidance provided in regulations and standards can play a key role in reducing incidents, but be careful not to fall into a compliance-only mindset (see Section 3.2). One successful example how an agency went beyond merely expecting compliance is the case of US OSHA providing supporting guidance on grain handling regulations (OSHA 2005).

Between 1976 and 1987, the USA averaged around 23 explosions in grain handling operations per year resulting in approximately 12 deaths and 40 injuries. In 1987 US OSHA promulgated a regulatory standard for grain handling (OSHA 1987) codifying many findings from past explosions. Table 2.2 shows that following implementation of the new regulation, the number of explosions was cut in half, and fatalities dropped by a factor of six. Then in 2005, US OSHA provided additional guidance on what they expected for compliance. This step amplified the positive impact of the regulations—incidents, fatalities, and injuries dropped even lower.

Table 2.2 Impact of US OSHA Grain Handling Standard (OSHA 2020)

Time Period	Explosions	Deaths	Injuries
1976-1987	~23/year	~12/year	~40/year
1987—New US OSHA regulatory standard for grain handling			
1988-2005	~12/year	~2/year	~13/year
2005—Additional US OSHA guidance			
2006- 2011	~8/year	~1/year	~5/year

A similar evolution has been seen with other regulatory and voluntary consensus standards. Dust explosion prevention is covered in the EU by the ATEX Directive (EU 2020) and by voluntary consensus standards in the USA (NFPA 2020), Brazil (ABNT 2018) and other parts of the world. This example reinforces the value of understanding regulations—even if they don't directly apply to you—and staying abreast of changes and new guidance.

More generally, whether or not your process is covered by a regulation or standard, you can and should take the opportunity to learn from these regulations and standards and apply what you learn. While compliance is always necessary, it is even more necessary to prevent process safety incidents.

2.2 Resources for Learning

You can learn about process safety incidents from a variety of resources that are available in various formats. This section will cover the resources most commonly used throughout the industry.

2.2.1 Process Safety Boards

Increasingly, countries are establishing process safety boards, quasi-independent governmental or non-governmental bodies that investigate incidents solely for the purpose of understanding root causes and communicating findings and recommendations broadly. Table 2.3 lists some of the better-known process safety boards. Some, like the US CSB and the Dutch Safety Board (DSB), also produce high-quality, engaging videos to enhance their communications. Nearly all the incident reports from these boards have been indexed in the Appendix.

Table 2.3 Examples of Established Process Safety Boards

Country	Board name
Brazil	Agencia Nacional do Petróleo
Japan	Association for the Study of Failure of Japan
Netherlands	Dutch Safety Board (DSB)
UK	COMAH Competent Authority of the Health and Safety Executive (HSE)
USA	Chemical Safety and Hazard Investigation Board (CSB)

2.2.2 Databases

In today's world, information is at your fingertips. Databases offer a large array of information that you can mine in a relatively short amount of time. Given the efforts to compile and maintain such databases, there may be a fee associated with their use. In addition to the databases developed by the process safety boards listed above, other commonly used process safety databases include:

- Analysis, Research, and Information on Accidents (ARIA 2020a)
- CCPS Process Safety Incident Database (CCPS 2020b)
- DECHEMA ProcessNet (DECHEMA 2020)
- Études de Prévention par l'Informatisation des Comptes Rendus d'Accidents (INRS 2020)

- European Commission Major Accident Reporting System (European Commission 2020)
- Infosis ZEMA (ZEMA 2020)
- Lessons Learned Database (IChemE 2020)
- Marsh 100 Largest Losses in the Hydrocarbon Industry (Marsh 2018).

2.2.3 Publications

Whether your bookshelf is on the wall, on a hard drive, in the cloud, or in a virtual or physical corporate library, you probably have a few books describing how to investigate incidents and extract lessons you can learn from them. Table 2.4 describes the most popular of these books. Additionally, most books about process safety use case histories featuring actual incidents to highlight key concepts.

Table 2.4 Books About Incidents and Incident Investigation

Author and/or Publisher	Name of book or series
CCPS/AIChE and Wiley	*Guidelines for Investigating Process Safety Incidents*, 3rd ed.
	Incidents That Define Process Safety
	More Incidents That Define Process Safety
Earl Boebert and James Blossom, Harvard University Press	*Deepwater Horizon: A Systems Analysis of the Macondo Disaster*
Andrew Hopkins, CCH Australia	*Failure to Learn: The BP Texas City Refinery Disaster*
	Lessons from Longford: The ESSO Gas Plant Explosion
Trevor Kletz, Butterworth Heinemann/IChemE	*What Went Wrong?* series
Frank Lees, Elsevier	*Loss Prevention in the Process Industries, eth ed.*
Roy Sanders, Butterworth Heinemann	*Chemical Process Safety Learning from Case Histories*, 3rd ed.

Newsletters and magazine articles can be a great way to quickly gather information. They are typically a short read with a big impact. Many websites contain archives of newsletters that allow the user to do keyword searches. For example, the CCPS *Process Safety Beacon* has over 70 keywords from which users can select to refine their search (Figure 2.3). Below is a list of the more prominent newsletters, magazines, and video series.

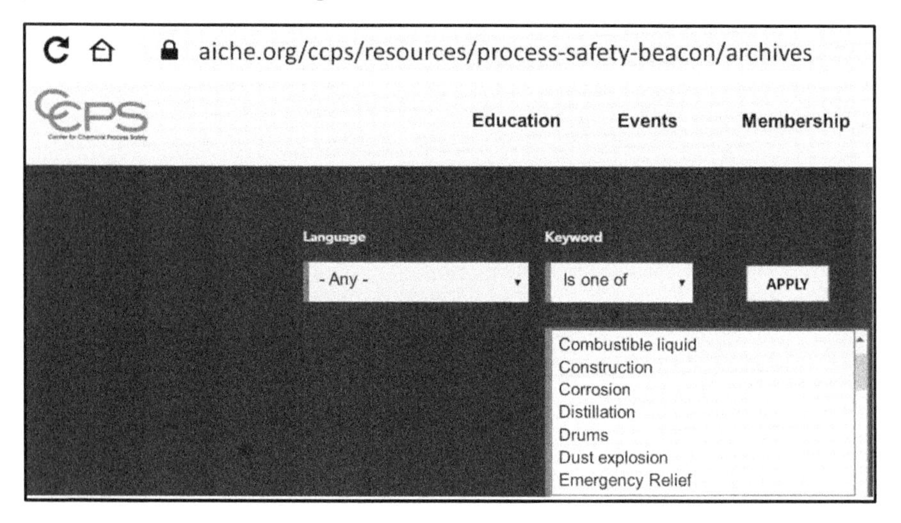

Figure 2.3 Process Safety Beacon Archive (Source: CCPS 2020c)

Newsletters

- *Process Safety Beacon* (CCPS 2020c)
- *Loss Prevention Bulletin* (IChemE 2020a)
- *EPSC Learning Sheets* (EPSC 2020)
- ICI Safety Newsletters by Trevor Kletz and Alan Rimmer (IChemE 2019)
- *Combustible Dust Newsletter* (DSS 2020).
- IOGP Safety Alerts (IOGP 2020)
- The BARPI (ARIA 2020b)

Journals and Magazines

- *Process Safety Progress* (AIChE 2020a)
- *Journal of Loss Prevention in the Process Industries* (Elsevier 2020).
- *Hindsight* (Eurocontrol 2020).

Video Series

- Napo: Safety with a Smile (EASHW 2020)
- Toolbox: Putting Safety in your Hands (EI 2020).
- Safety Lore (IChemE 2020b)

2.2.4 Events and Proceedings

Learning opportunities abound at technical conferences, symposia, and workshops. The traditional learning method is to listen to knowledge shared during a presentation. You can learn so much more, however, by engaging with the presenter. A presenter can only give so much information in a brief presentation. Taking the opportunity to interact one-on-one with the presenter allows you to ask questions that are critical for your specific situation.

Similarly, take advantage of networking events and local and national trade associations to build trusted relationships with new colleagues. There are also ample opportunities to build relationships by participating in standards committees and or joining process-safety focused organizations like CCPS. Collaboration helps you and your colleagues build each other's knowledge bases. In fact, "Responsible Collaboration" is a fundamental tenet of CCPS's Vision 20/20, a roadmap to how industry could attain and maintain excellence in process safety (CCPS 2020d).

As a corollary to attending technical events, conference and symposium proceedings are a useful resource for learning about external incidents that may not be covered in incident databases. Even if you can't attend the conference, the published proceedings often provide you with the presenters' contact information so that you can contact them with questions. Examples of events with proceedings include:

- AIChE/CCPS Global Congress on Process Safety (AIChE 2020b), especially the "Case Histories" sessions
- AIChE Ethylene Producers' Committee Conference (AIChE 2020b, 2020c), especially the safety sessions
- AIChE Safety in Ammonia Plants and Related Facilities Symposium (AIChE 2020d)
- Other CCPS Global and Regional Process Safety Conferences in China, Europe, India, Japan, Latin America, Mideast, and elsewhere (CCPS 2020e)

- American Fuel and & Petrochemicals Manufacturers Conferences (AFPM 2020).
- IChemE Hazards Symposium (IChemE 2020c)
- International Symposium on Loss Prevention and Safety Promotion in the Process Industries (EFCE 2019).

2.2.5 Other Resources

Many other resources can help you learn from incidents. Consulting companies, insurers, and manufacturers will, with varying degrees of anonymity, publish papers in journals and conferences describing incident investigations and their findings and recommendations. When you attend regularly scheduled process-safety training, you not only keep your skills up to date, but often you can get a closer look at specific incidents.

Virtual events allow you to glean information across the globe—without having to travel. Virtual learning opportunities include webinars, e-Learning modules, and micro-learning opportunities. We can also anticipate advances in virtual immersive learning experiences with technologies such as virtual- and augmented-reality.

2.3 References

2.1 Associação Brasileira de Normas Técnicas (2018). *Atmosferas Explosivas*. Place of publication: Associação Brasileira de Normas Técnicas.

2.2 AFPM (2020). *Technical Papers*. www.afpm.org/data-reports/technical-papers (accessed May 2020).

2.3 AIChE (2020a). *Process Safety Progress*. AIChE/Wiley, aiche.onlinelibrary.wiley.com/journal 15475913 (accessed May 2020).

2.4 AIChE (2020b). Proceedings of the AIChE Spring Meeting and Global Congress on Process Safety. New York: AIChE.

2.5 AIChE (2020c). Proceedings of the Ethylene Producers' Conferences. epc.omnibooksonline.com (accessed May 2020).

2.6 AIChE (2020c). Proceedings of the Safety in Ammonia Plants and Related Facilities Symposium. waiche.org/ammonia (accessed May 2020).

2.7 ARIA (2020). The ARIA Database. www.aria.developpement-durable.gouv.fr/the-barpi/the-aria-database/?lang=en (accessed June 2020).

2.8 ARIA (2020). The BARPI. Subscription via www.aria.developpement-durable.gouv.fr/the-barpi/?lang=en (accessed June 2020).

2.9 CCPS (2007). *Guidelines for Risk Based Process Safety*. Hoboken, NJ: AIChE/Wiley.

2.10 CCPS (2020a). CCPS and Process Safety Publications. www.aiche.org/ccps/publications#books (accessed May 2020).

2.11 CCPS (2020b). PSID: Process Safety Incident Database. www.aiche.org/ccps/resources/psid-process-safety-incident-database (accessed May 2020).

2.12 CCPS (2020c). Process Safety Beacon Archives. www.aiche.org/ccps/resources/process-safety-beacon/archives (accessed May 2020).

2.13 CCPS (2020d). Moving from Good to Great: Guidelines for Implementing Vision 20/20 Tenets & Themes. www.aiche.org/ccps/moving-good-great-guidelines-implementing-vision-2020-tenets-themes (accessed June 2020).

2.14 CCPS (2020e). Conferences. www.aiche.org/ccps/resources/ conferences (accessed June 2020).

2.15 CSB (2006). Combustible Dust Hazard Investigation. CSB Report No. 2006-H-01.

2.16 CSB (2016). Combustible Dust Safety. www.csb.gov/recommendations/mostwanted/combustibledust. (Accessed January 2020).

2.17 CSB (2017a). ExxonMobil Refinery Chemical Release and Fire. CSB Report No. 2016-02-I-LA.

2.18 CSB (2017b) Didion Milling Company Explosion and Fire. www.csb.gov/didion-milling-company-explosion-and-fire-/.

2.19 DECHEMA (2020). DECHEMA ProcessNet. processnet.org (accessed June 2020).

2.20 DSS (2020). Resources. dustsafetyscience.com/resources (accessed June 2020).

2.21 EASHW (2020). Napo: Safety with a Smile. osha.europa.eu/en/tools-and-resources/napo-safety-smile (accessed June 2020).

2.22 EFCE (2019). International Symposium on Loss Prevention and Safety Promotion in the Process Industries. lossprevention2019.org/ (accessed June 2020).

2.23 EI (2020). Toolbox: Putting Safety in your Hands. toolbox.energyinst.org (accessed June 2020).

2.24 Elsevier (2020). *Journal of Loss Prevention in the Process Industries*. Amsterdam: Elsevier.

2.25 European Commission (2020). European Commission Major Accident Reporting System. emars.jrc.ec.europa.eu (accessed June 2020).

2.26 EPSC (2020). Learning Sheets. epsc.be/ Learning+Sheets.html (accessed June 2020).

2.27 EU (2020). Equipment for potentially explosive atmospheres (ATEX). ec.europa.eu/growth/sectors/mechanical-engineering/atex_en (accessed May 2020).

2.28 Eurocontrol (2020). HindSight. www.skybrary.aero/index.php/HindSight_-_EUROCONTROL (accessed June 2020).

2.29 HSE(1994). *The Fire at Hickson & Welch Limited. A report of the investigation by the Health and Safety Executive into the fatal fire at Hickson & Welch Limited, Castleford on 21 September 1992.* London, UK: Her Majesty's Stationary Office.

2.30 IChemE (2020). ICI Newsletters. www.icheme.org/membership/communities/special-interest-groups/safety-and-loss-prevention/resources/ici-newsletters (accessed December 2019).

2.31 IChemE (2020a). Loss Prevention Bulletin. www.icheme.org/knowledge/loss-prevention-bulletin (accessed month year).

2.32 IChemE (2020b). Safety Lore. www.icheme.org/knowledge/safety-centre/safety-lore (accessed June 2020).

2.33 IChemE (2020c). Hazards. www.icheme.org/hazards29 (accessed June 2020).

2.34 IChemE (2020d). Lessons Learned Database. www.icheme.org/membership/communities/special-interest-groups/safety-and-loss-prevention/resources/lessons-learned-database (accessed July 2020).

2.35 INRS (2020). EPICEA (Études de Prévention par l'Informatisation des Comptes Rendus d'Accidents). www.inrs.fr/publications/bdd/epicea.html (accessed June 2020).

2.36 IOGP (2020). Safety Zone: Safety Alerts. safetyzone.iogp.org/safetyalerts/alerts (accessed April 2020).

2.37 Marsh (2020). *The 100 Largest Losses 1978–2019: Large Property Damage Losses in the Hydrocarbon Industry*, 26th ed. New York: Marsh & McLennan Companies.

2.38 NFPA 654 (2020). *Standard for the Prevention of Fire and Dust Explosions from the Manufacturing, Processing, and Handling of Combustible Particulate Solids*. Quincy, MA: National Fire Protection Agency.

2.39 OSHA 1910.272 (1987). *Grain Handling Facilities Standard*. Washington, DC: US Occupational Safety and Health Administration.

2.40 OSHA (2005). Grain Handling. US Occupational Safety and Health Administration. www.osha.gov/SLTC/grainhandling (Accessed March 2020)

2.41 OSHA (2020) Grain Elevator Explosion Chart. www.osha.gov/SLTC/grainhandling/ explosionchart.html (accessed March 2020).

2.42 ZEMA (2020). Infosis ZEMA. www.infosis.uba.de/index.php/en/site/13947/zema (accessed April 2020).

3
OBSTACLES TO LEARNING

"Once you stop learning, you start dying."
—Albert Einstein, Nobel Prize winning theoretical physicist

There's been an incident! Once it's safe to start the investigation, CCPS describes the incident investigation process (CCPS 2019a) as including the following steps:

- Gather evidence.
- Analyze the evidence.
- Determine the causal factors.
- Determine the root causes.
- Develop and issue recommendations that are achievable and will lead to improvement.
- Implement recommendations and ensure follow-up.

A successful investigation—defined as one that leads to lasting improvement in the way the facility or company operates—depends on successful, professional completion of the investigation by competent investigators. Although the technical details considered in each step can be complex, the high-level process is simple enough. Unfortunately, organizational obstacles routinely interfere. Sometimes, getting to learnings can be a challenge, as human nature dictates that no one wants to be at fault. When blame-shedding interferes with your investigation, think of what Jim Collins wrote in his landmark management book *Good to Great* (Collins 2001):

> You must maintain unwavering faith that you can, and will, prevail in the end, regardless of the difficulties, and at the same time, have the discipline to confront the most brutal facts of your current reality, whatever they might be.

In this section we examine some of the most common obstacles to learning from past incidents (Table 3.1). A more detailed discussion of obstacles to institutional learning was described by Schilling and Kluge (Schilling 2009). Keep in mind that not all companies experience every obstacle, and companies may overcome obstacles in the future; meanwhile new ones may arise. It is important to continuously assess the corporate learning process and to address obstacles as they appear.

Table 3.1 Common Obstacles to Individual and Company Learning

Individual	Company
Organizational changes	Cost and business pressures
Retirements and job changes	Reverse incentives
Natural memory loss and normalization of deviance	Leaner organizations
Insufficient or incomplete evaluation of hazards	Risk misperception
Lack of understanding about hazards	Compliance-only mindset and over-anticipation of litigation
Difficulty to see beyond own experience or type of industry	Too many high priorities or rapidly changing priorities
Both	
It can't happen here attitude—loss of sense of vulnerability	
Ivory tower syndrome or lack of communication	
Assessing blame rather than correcting root causes	
Misplaced conservatism	

3.1 The Impact of Individuals

In any organization, the personnel roster is in constant flux. People change roles, get promoted, or leave the company for other opportunities. Organizational change can impact corporate memory in several ways. Almost by definition, each organizational change results in an incumbent being replaced by someone with less organizational memory related to that position. Unless the organizational memory has been captured in the company's PSMS, standards, policies, and culture, memory will be continuously lost.

Organizational Changes

With management turnover, the organization's priorities can shift. New leaders may emphasize different goals than their predecessors or add new goals. A shift in goals might directly impact the way the organization practices process safety or might indirectly shift attention away from process safety. Companies must rigorously train new employees to ensure that they have learned what they must know about process safety and will maintain continuous improvement.

Retirements

Currently, another form of organizational change is being driven by demographics. Every retirement creates a situation in which an individual with significant experience must pass knowledge to his or her successor. This transfer of knowledge is hard enough when an experienced successor is available, but the current wave of baby-boom retirements is especially challenging because experienced replacements are in short supply. An economic turndown in the 1980s and 1990s drove new engineering graduates to enter other fields, such as finance, medicine, and law. The result is a smaller pool of engineers with the institutional knowledge needed to replace the boomers. While some of a retiring employee's knowledge can be captured through documentation, the lack of hands-on training may result in the loss of corporate memory.

Natural Memory Loss and the Quest for Improvement

Studies (Throness 2013; Ebbinghaus, 1885) have shown that in the absence of periodic reminders, people and companies generally forget incident lessons learned within three years, and possibly sooner (see also Chapter 7). Conversely, humans naturally seek to improve on established procedures if they sense a potential to innovate and improve efficiency. If improvements are not carefully managed, they can introduce new risks and undo important lessons learned. Fortunately, if people are reminded regularly of an incident and its findings, and if the learnings are embedded in the culture, this knowledge can stay indefinitely.

Insufficient or Incomplete Evaluation of Hazards

In 1966, a fire swept through the Apollo 1 capsule during a "plugs-out" test on the launchpad, killing all three astronauts. Before the test, during the hazard/risk analysis, NASA did not recognize problems with the

See Appendix index entry S8

capsule design. The original design featured an oxygen-rich atmosphere, lots of nylon straps, and an inward-opening hatch. During the test, an electrical short occurred. Nylon normally smolders in air, but the straps combusted rapidly in the capsule's high-oxygen atmosphere. Once the fire started, the pressure from the combustion gases tightly sealed the hatch. Despite the efforts of the crew inside the capsule and ground support staff outside, the hatch could not be opened. It was impossible for the astronauts to escape.

The investigation committee called this breakdown of the hazard analysis a failure of imagination. However, the astronauts themselves imagined it. During a planning meeting, they asked that flammables be removed from the Apollo cabin. But the designers gave this step a low priority, partly because NASA considered a cabin fire improbable. The astronauts also were said to have asked for an outward-opening door.

The Apollo 1 fire is an example of the natural human tendency to understate potential consequences, understate probability of occurrence, and overstate the effectiveness of preventive and mitigative barriers. These tendencies become exacerbated when hazard and risk analysis is performed under time pressure with competing priorities.

A lesson learned that became institutional knowledge

In May 2020, SpaceX and NASA successfully launched the Crew Dragon spacecraft to the International Space Station. At the pre-launch briefing, Astronaut Doug Hurley said, "On more than one occasion he [Elon Musk] has looked both Bob [Astronaut Behnken] and me right in the eye and said, 'Hey, if there's anything you guys are not comfortable with or that you're seeing, please tell me and we'll fix it.'" (Grush 2020). This appears to be a step in the right direction for institutionalizing past lessons learned. Now NASA and its suppliers need to ensure this lesson remains institutionalized.

Lack of Understanding About Hazards

Sometimes, you don't know to look for a hazard until the hazard finds you. The 1960 aniline plant explosion in Kingsport, TN, is one good example. In the presentation "Let Me Tell You...The Impact of Eastman's Aniline Plant Explosion on Process Safety Awareness" at the 11th Global Congress on Process Safety (Lodal 2015), the speaker noted that at the time of the 1960 investigation, the cause of the explosive reaction was unknown. To this day, the cause is still only speculative, but it has spurred on greater awareness that process safety hazards can exist in unexpected places.

Difficulty to See Beyond Own Experience or Type of Industry

As mentioned in the previous chapter, we often do not take advantage of learnings from incidents in other industries that may well be applicable to our own. What's more, given the time pressures of today's workload, we often rely on the experiences that we are familiar with and may decide not to look any further. Limiting knowledge to a specific scope limits the possibility of predicting or perhaps even preventing incidents.

Distraction from Lower Profile Hazards by High-Profile Hazards

Refinery personnel generally recognize and strive to control flammable hazards, the special hazards of hydrogen, and the toxic hazards of H_2S and HF. Lesser hazards such as combustible coke and sulfur dust may receive less attention, however. In one refinery fire, cylinders of chlorine gas used to treat wastewater were compromised. These cylinders were never considered in HIRA.

The Bhopal plant handled phosgene, a well-known toxic chemical that had been used as a chemical weapon. Designers ensured that phosgene vessels were as small as possible, so that if a release occurred, the consequences would be reduced. But the more toxic MIC was stored in three large tanks.

3.2 The Impact of Company Culture

Another common obstacle to learning is the culture of the organization. "Time is money" is a common phrase we hear. This line of thought can lead to an unbalanced view of risk, so that process safety risk takes a back seat to company profits. It can also result in a culture of "If it ain't broke, don't fix it," leading to reactive approaches rather than proactive ones.

Cost and Business Pressures

Anyone who has spent time in a process safety leadership role has probably heard a senior manager say, "Every time you come into my office it costs me another million dollars." Many requests for additional spending on safety come from findings from external and internal incidents and a genuine wish to apply these improvements plant- or company-wide.

Even when safety improvements do not come with a capital cost, process safety professionals may find themselves being blamed for holding up process modifications for management of change (MOC) reviews and slowing production for scheduled maintenance, inspection, and testing. Over time, cost and business pressures can make less thick-skinned individuals reluctant to seek knowledge from past incidents and instead accept the status quo.

Reverse Incentives

The best process safety management systems use metrics to track performance and guide improvement. It can be tempting to also use metrics to offer leaders and workers incentives towards better process safety performance. However, incentives can have reverse effects if they are not carefully designed.

For example, imagine that a plant had the goal of reducing incidents by closing the PSMS gaps that led to near-misses. If personnel were given an incentive based on decreasing the number of near-misses, it could drive them to hide near-misses. This would almost certainly lead to increased incidents.

Similarly, the plant might have a goal to increase the closure of action items from process hazard analyses (PHAs), audits, incident investigations, and so on. Closure of action items should reduce the incidence of both near-misses and incidents. However, unless the plant also implements a system for verifying that action items are closed properly, personnel could be tempted to check the box rather than close action items with effective measures.

Leaner Organizations

In the continuing effort to improve efficiency, plants and corporate groups continue to grow leaner. Companies expect fewer personnel to cover a broader range of responsibilities, often aided by automation. Often, as organizations grow, the existing personnel must cover expanded responsibilities. This leaves personnel with less time to study the literature and think strategically about improvement. Even when they have time, personnel may be reluctant to consider improvements, especially if it will mean additional work for them. If the company wishes to continuously learn from process safety incidents, it must provide key personnel the time they need to study past incidents and develop ways to integrate those learnings into the way the company operates.

Risk Misperception

The risk of process safety incidents is driven by low probabilities and high consequences. Most other risks that companies manage, including occupational safety, have higher probabilities but relatively lower consequences. Process safety incidents are relatively rarer than other adverse events. This can lure leaders and employees alike into thinking that process safety risks are lower than they really are, or that the process safety problem has been solved.

This misperception can be further enhanced if the company is improving performance in occupational safety. Despite their common use of the word safety (and a few areas of overlap), process safety and occupational safety differ significantly in their general competencies, technologies, and management structures. Yet improved performance in occupational safety can lead to intentional or unintentional de-prioritization of process safety, de-emphasizing continuous learning.

Risk misperception can also occur when companies have previously had negative HSE experiences with one chemical—or even in a single HSE area. As a result of past experiences, the perceived risk seems higher than it might have otherwise. A company with past landfill liability may view managing its landfills as riskier than any process safety hazard and therefore divert process safety resources. Similarly, a company that has had many workers get cancer from occupational exposure to a chemical may manage that chemical as toxic—and only toxic— although it is also flammable and reactive.

Ultimately, if a practice is considered less important, it's easy to just forget about it. Make it your goal to ensure that the importance of maintaining and continuing to improve process safety becomes deeply embedded in the corporate culture—so that these misperceptions can never occur.

Compliance Mindset and Anticipation of Litigation

CCPS (CCPS 2019b) has described the perils of relying on the compliance-only mindset:

- Regulations are designed to protect society, not to protect the company.
- Regulations don't cover everything you need to do.
- Regulators inspect only rarely.
- Regulators don't know your process as well as you do.

A finding of compliance by a regulatory agency is no guarantee that you are managing your process safety risks adequately. However, such a finding may fool people into thinking there is no need for improvement. In this kind of environment, there is no incentive for learning.

A compliance-only mindset often goes together with a mindset of continuously expecting litigation. The legal outcomes of both criminal and civil litigation may be worse if the company knows of hazards or improvement opportunities but fails to address them. Some attorneys and managers discourage continuous learning for this reason.

This approach virtually guarantees eventual litigation, however. Ultimately, the unknown, unresolved problems become incidents that draw the regulators' attention. It is much better to seek knowledge and address gaps so that there are fewer incidents requiring legal defense.

Many regulations and standards will consider processes and equipment that were designed to an earlier version of a standard current at the time to be in compliance even if the standard is later changed. This acceptance of legacied designs can allow a plant to be in full compliance, but not meet the company's risk criteria. As CCPS discusses in Vision 20/20 (the organization's guiding vision for process safety by the year 2020), it is important to monitor standards for changes and determine if process or equipment improvements are needed for legacy-compliant designs (CCPS 2014).

3.3 Obstacles Common to Individuals and Companies

"It Can't Happen Here" Attitude—Loss of the Sense of Vulnerability

A sense of vulnerability is an essential characteristic of a good process safety culture. We all know the importance of maintaining a healthy respect for process hazards and using that as motivation to faithfully execute our roles with professionalism. Catastrophic incidents are infrequent, however, and that can drive us to relax our sense of vulnerability, leading to complacency and a false sense of security—which in turn can compromise performance and demotivate efforts to improve.

Ivory Tower Syndrome

Many companies have teams of highly competent process safety professionals focused on advanced learning. Often, however, these individuals are effectively walled off from both operations and corporate oversight roles. This can create a significant gap between what the corporate experts have learned and what the company practices.

In some cases, the walls are real organizational obstacles, while in other cases they are built by the personalities of the individuals involved. In either case, the company cannot benefit from what its experts have learned. To obtain the maximum learning benefit from these experts, companies should institute systems to transfer the experts' knowledge into the corporate memory.

Assessing Blame Rather than Correcting Root Causes

Across industries, we continue to struggle to perform in-depth incident investigations that identify root causes that can be corrected. Investigators often stop once they identify the individual who made the ultimate error and correct the error by punishing the individual.

Although punishment could potentially prevent the individual from making the same error, this step doesn't address systematic process or design problems or work conditions that led to the error. That makes the error likely to happen again. Worse, it may drive others in the organization to hide incidents and near-misses, preventing the company from being able to fix the root cause.

The desire to avoid blame may also lead managers to limit the scope of an incident investigation or improvement recommendations. This action may be driven by legal counsel seeking to avoid fines or lawsuits or by other personnel fearing a negative performance review.

Misplaced Conservatism

One of the ironies of the human condition is that as much as we seek to continuously improve, innovate, and advance technology, we also resist change, even when change is well justified. Both extremes—progressivism and conservatism—can create problems in process safety. In plants and companies with a culture of doing business as usual, opportunities to improve from learning may be unnecessarily blocked.

Most of the factors described in this section can lead to gradual degradation of culture and corporate memory. This outcome can only be combatted by seeking continuous improvement. This will help ensure that any institutional knowledge that may have briefly been forgotten will become reinstated within the culture.

3.4 Consequences of Not Learning from Incidents

When we don't learn from incidents, we run a higher risk of repeating them. Learning from incidents is important because human lives are at stake. Incidents can result in fatalities and/or injuries to employees or to the public. They can result in property damage to the plant and the surrounding area.

They can result in the release of harmful materials. Ultimately, they cause reputational damage and break the bond of trust, whether between employee and employer or between the company and the public.

No company wants to have an incident memorialized in a blockbuster Hollywood movie the way that BP did with the Macondo Deepwater Horizon blowout (Berg 2016), an enduring reminder of the consequences of not learning from incidents. On a lesser scale, but still no better, no one should want to make it into National Geographic's *Seconds from Disaster* series, which covered the 2005 Texas City, TX, USA, incident; Bhopal, India; Fukushima Daiichi, Japan; and Chernobyl, Russia (National Geographic 2020). A quick search on YouTube using the keywords "process safety incident" yields a slew of CSB animations of incidents, and a keyword search on "Pemex Reynosa Explosion" shows us what happened in real time, the horrifying result of not learning from past incidents (Reynosa 2012).

To maintain a solid reputation as a good neighbor, companies must maintain a culture of trust and open, frank communications and commit to understanding and acting on hazards and risks well before they become an incident or serious near-miss. These interactions must continue over time, or the trust and communication will diminish. If organizations lose the public's trust, it will be difficult to regain and can lead to resistance to change (such as expansions) in the future (CCPS 2019b).

3.5 References

3.1 Berg, P. (director) (2016). *Deepwater Horizon*. United States: Summit Entertainment (Lionsgate).

3.2 CCPS (2014). Vision 20/20 Process Safety: The Journey Continues. New York: AIChE.

3.3 CCPS (2019a). *Guidelines for Investigating Process Safety Incidents*, 3rd Ed. Hoboken, NJ: AIChE/Wiley.

3.4 CCPS (2019b). *Process Safety Leadership from the Boardroom to the Frontline*. Hoboken, NJ: AIChE/Wiley.

3.5 CSB (2007). Valero Refinery Propane Fire. CSB Report No. REPORT NO. 2007-05-I-TX.

3.6 Ebbinghaus, H. (1885). *Memory: A Contribution to Experimental Psychology*. New York: Dover.

3.7 Grush, L. (2020). Meet the first NASA astronauts SpaceX will launch into orbit. www.theverge.com/2020/5/20/21254315/spacex-crew-dragon-astronauts-behnken-hurley-nasa-launch-dm-2. (Accessed 20 May 2020).

3.8 Lodal, P.N. (2015). Let me tell you...The impact of Eastman's aniline plant explosion on process safety awareness. *Proceedings of the 2015 AIChE Spring Meeting and 11th Global Congress on Process Safety,* Austin, TX (26-30 April 2015). New York: AIChE.

3.9 National Geographic (2020). *Seconds from Disaster*. www.natgeotv.com/za/shows/natgeo/seconds-from-disaster (accessed April 2020).

3.10 Reynosa (2012). Gas Plant Explosion Mexico [video]. www.youtube.com/watch?v=6jhCKp2LHro (accessed April 2020).

3.11 Schilling, J and Kluge, A. (2009). Barriers to organizational learning: An integration of theory and research. *International Journal of Management Reviews* 11 (3): 337–360.

3.12 Throness, B. (2013). Keeping the memory alive, preventing memory loss that contributes to process safety events. *Proceedings of the Global Congress on Process Safety*, San Antonio, TX (28 April–2 May 2013). New York: AIChE.

4
Examples of Failure to Learn

"The past can't hurt you anymore, not unless you let it."
—*From* V is for Vendetta *by Alan Moore*

The Index of Publicly Evaluated Incidents prepared for this book (see Appendix) illustrates how often companies fail to learn from past incidents. Figure 4.1 shows that of the 441 incidents in the index, 13 primary and secondary findings were repeated in more than 100 incidents. (Primary findings are those that contributed most to the incident; secondary findings are those that contributed less but still may have learning potential.) Three of these findings—Hazard Identification and Risk Analysis (HIRA), Culture, and Safe Design—repeated in more than 180 incidents. As large as they are, these numbers may be understated, since the indexing process intentionally limited the number of findings that would be indexed for each incident.

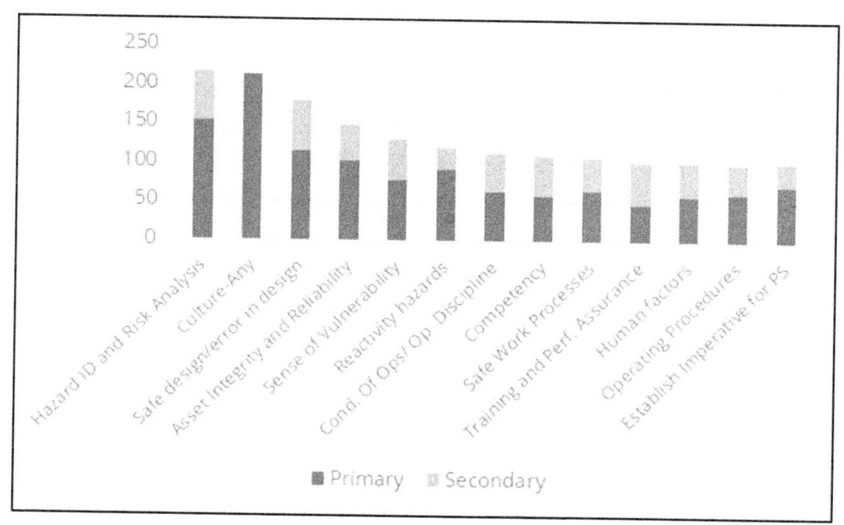

Figure 4.1 Most Repeated Findings from 441 Indexed Incidents

You may not have heard of most of the incidents in this index, but you should have heard of the major disasters described in this chapter and Chapter 8 (see Table 8.1 for a summary of landmark incidents with their associated prominent findings and causal factors). You might expect that the most well-known incidents had unusual causes, but it's clear that many had the same causes as incidents that preceded them. Each of these incidents has been studied deeply, with the findings published, distributed broadly, and pored over by safety professionals around the world. Nonetheless, the causes of the most prominent incidents are repeated in subsequent incidents. In this chapter, we examine a few of these repeating failures in more depth.

4.1 Process Safety Culture

Process safety culture is described by CCPS (CCPS 2018) as:

> ...the combination of group values and behaviors that determine the manner in which process safety is managed. It's how we conduct ourselves when we think no one is watching.

An effective and sound process safety culture is a foundational element in CCPS's Risk Based Process Safety (RBPS) pillar of Committing to Process Safety. This culture must be established at all levels, from the boardroom to the front line. Everyone in the organization has a role in process safety culture and must perform their role reliably and with professionalism.

A weak process safety culture often undermines successful execution of the other PSMS elements, increasing the probability of an incident. One commonality among the landmark incidents covered in Chapter 8 is that they all had a weak process safety culture. Although these incidents have been well documented, companies are not learning from them, suggesting a breakdown in process safety culture.]

See Appendix index entry C26

In the 2014 LaPorte, TX, USA, incident, the primary focus of the company's safety culture program was personal safety. This helped the company to reduce its US OSHA total recordable injury rate. But according to the CSB report on this incident (CSB 2019), the company never evaluated its process safety culture. The report stated:

> Had its efforts included a focus on perceptions of process safety as well as personal safety in its Safety Perception Survey, or had it performed a separate process safety culture assessment with the intent of improving

process safety culture as recognized in the company's corporate process safety management standard, the company likely would have been more aware of potential process safety issues and better positioned to prevent or mitigate future process safety incidents.

Process safety culture breakdowns lead to poor decision-making, especially under the pressure to produce. This is often true outside as well as inside the process industries. The 2018 and 2019 incidents involving the crash of two 737 Max airplanes (resulting in 346 fatalities) had some eerie similarities to the 1986 Space Shuttle *Challenger* incident, where the engineers' recommendation to delay the launch was overruled by management. At a House Transportation Committee hearing, retired aerospace engineer G. Michael Collins testified that "a total of at least thirteen Federal Aviation Administration (FAA) aerospace engineers, one pilot, and at least four FAA managers disagreed" with the manufacturer's position that the planes were safe to fly. In an interview with the *Washington Post*, Collins went further, saying, "The existing FAA management safety culture is broken and demoralizing to dedicated safety professionals" (Laris 2019)."

Despite the objections, FAA regulators essentially deferred to the operating company. As Collins testified, FAA management culture has shifted to where the wants of applicants now often take precedence over the safety of the traveling public.

A formal report on the 737 Max incidents has not been issued at the time of this writing. However, preliminary reports link the cause of the crashes to a new automated flight control feature, the Maneuvering Characteristics Augmentation System, which repeatedly forced the aircraft to nosedive. Pilots and industry experts criticized the plane manufacturer for omitting information about this system from flight manuals and training materials.

See Appendix index entry S16

Another culture failure with many parallels to process safety is the 2019 Brumadinho dam collapse in Brazil. The collapse unleashed more than 12 million cubic meters of mining waste, causing hundreds of fatalities. Following the incident, the dam operator hired an expert panel to investigate (Robertson 2019a, 2019b). The panel found that the collapse of the dam was years in the making, driven by a combination of internal strains due to creep and strength reduction due to loss of suction in the unsaturated zone because of intense rainfalls. The panel determined that the original design and construction of the dam was flawed in two ways: it retained water instead of allowing it to drain, and its slope was very steep, placing increased stress on the material in the dam.

This problem was recognized at least as early as the year before the collapse. Brazilian authorities required an independent expert to audit the dam annually and sign off that the dam remained safe. At the September 2018 audit, however, the independent auditor refused to sign, citing the deficiencies later identified in the investigation report. Not deferring to the expertise of the auditor, the mine hired a second auditing company. Following the incident, representatives of the second company testified that they were pressured by the dam operator to sign off (Quintão 2019).

4.2 Facility Siting

Facility siting is defined as an assessment of the potential damage to occupied buildings from explosions, fire, and toxic hazards. A facilities siting assessment includes not only the buildings on the plant site, but also residential and commercial buildings surrounding the site. Many landmark incidents, including Bhopal, MP, India in 1984, Texas City, TX, USA in 2005, and West, TX, USA in 2013 incidents had significant facility siting issues.

Facility Siting—Offsite Consequences

The facility siting issue at Bhopal in 1984 pertained to tin shacks erected by squatters adjacent to the plant. Both the government and the plant intended for this area to be kept unoccupied but failed to enforce that intention.

Not long before the Bhopal tragedy, a Mexico City fuel terminal consisting of gas storage spheres and bullet tanks was destroyed in a massive series of explosions. The cause of the incident was a burst pipeline releasing a vapor cloud that was ignited by a flare used to burn off excess gas; shortly thereafter, a boiling liquid expanding vapor explosion (BLEVE) occurred, followed by 15 more such explosions over the next 90 minutes. The bursting vessels became projectiles that flew more than 100 meters. The explosion destroyed the entire facility, resulting in the deaths of all workers. It also caused major offsite damage and more than 500 fatalities in the community. According to a summary by the Health and Safety Executive (HSE) of the UK (HSE 2020):

> The total destruction of the terminal occurred because there was a failure of the overall basis of safety, which included the layout of the plant and emergency isolation features.

The HSE summary goes on to explain that the plant layout did not factor in adjacent hazards that resulted in a domino effect when the first BLEVE event occurred. The fire-water system was disabled after the initial blast and the water-spray system was inadequate, allowing for the fire to become uncontrollable.

The lessons about facility siting from Bhopal and Mexico City had still not been learned nearly 30 years later, as the West, TX, USA, incident in 2013 shows. The West explosion severely impacted homes, businesses, schools, and a senior residence facility that had been built near the facility without recognizing or analyzing the potential explosion hazards. Some buildings were less than 150 meters (500 feet) from the facility.

And the lesson still hasn't been learned today. In May 2020, a styrene release incident in Vishkapatnam, AP, India, produced a toxic cloud that caused at least 11 fatalities in the nearby community and injured hundreds (Doyle 2020). As of this writing, the investigation is underway, but similar facility siting deficiencies to Bhopal and Mexico City may well be one of the findings.

Facility Siting—Onsite Consequences

In the Flixborough explosion (see Section 8.1 for more details), occupied buildings onsite, including the control room, were destroyed because they were too close to the process and not blast resistant (UKDOE 1975). The lessons from Flixborough had not been learned five years later, when another explosion at a high-density polyethylene plant in Pasadena, TX, USA, caused 23 fatalities and injured hundreds. While workers were clearing blockage of a settling leg, 39,000 kg (85,000 lb.) of flammable process gases were released. In the highly confined process rack, the gases ignited in a vapor cloud explosion. Two 76 M^3 (20,000-gal) isobutane storage tanks also exploded. Debris from the explosions spread over a six-mile radius.

The investigation into the Pasadena incident was conducted by US OSHA. Its findings in the April 1990 report to the President noted:

> The large number of fatally injured personnel was due in part to the inadequate separation between buildings in the complex. The distances between process equipment were in violation of accepted engineering practices and did not allow personnel to leave the polyethylene plants safely during the initial vapor release; nor was there enough separation between the reactors and the control room to carry out emergency shutdown procedures. The control room, in fact, was destroyed by the

initial explosion. Of the 22 victims' bodies that were recovered..., all were located within 250 feet (76 meters) of the vapor release point; 15 of them were within 150 feet (46 meters). (Dole 1990)

The lesson about onsite facility siting still had not been learned in the Texas City, TX, USA, incident in 2005 (see Section 8.4), where temporary trailers that should not have been located near an operating unit were destroyed, resulting in the deaths of 15 occupants.

And onsite facility siting lessons still haven't been learned today. In 2019, a leak in a mixing line on a naphtha tank in a Deer Park, TX, USA storage terminal led to a fire that caused 13 tanks to fail and a pipe bridge to collapse. The tanks and pipe bridge were inside a common dike, allowing the burning naphtha to spread around most of the dike (CSB 2019).

Facility siting standards

The American Petroleum Institute (API) has incorporated learnings from all these incidents and more in issuing several recommended practices including: RP752, Management of Hazards Associated with Location of Process Plant Permanent Buildings (API 2009); RP753, Management of Hazards Associated with Location of Process Plant Portable Buildings (API 2007); and RP756, Management of Hazards Associated with Location of Process Plant Tents (API 2014). Publishing these recommended practices was just the first step, capturing the learning. It is now up to all of us to ensure that this learning becomes embedded in our cultures.

4.3 Maintenance of Barriers/Barrier Integrity

Barriers are measures that reduce the probability of releasing a hazard. As Table 4.1 shows, there are two types of barriers: preventive barriers that prevent process materials from being released, and mitigating barriers that reduce consequences if a release occurs.

Table 4.1 Examples of Barriers for Controlling Process Safety Hazards

Preventive Barriers	Mitigating Barriers
Interlock	Dike or berm
Safety instrumented system	Emergency flare or scrubber
Alarm with prescribed action	Deluge or sprinkler system
Emergency cooling	Fireproofing
Procedural controls	Pressure relief system

Companies throughout the industry continue to forget that if barriers are not adequately maintained, the incidents they are designed to prevent can happen. Bhopal, MP, India, and Buncefield, Hertfordshire, UK, two landmark incidents where barrier integrity was compromised, will be discussed in detail in Chapter 8.

Storage tank overflow incidents are a recurring example of failure to learn about barrier maintenance. Table 4.2 provides a sampling of these incidents.

Table 4.2 Storage Tank Overflow Incidents

Year	Location	Material
1983	Newark, NJ, USA (CSB 2015)	Gasoline
1988	Yamakita, Kanagawa, Japan (ASF)	Hydrogen peroxide
2005	Buncefield, Hertfordshire, UK (HSE,2011)	Gasoline
2008	Petrolia, PA, USA (CSB 2009a)	Oleum
2009	Bayamón, PR, USA (CSB,2015)	Gasoline
2011	Reichstatt, Bas-Rhin, France (ARIA 2013)	Gasoline
2014	Fukushima, Japan (BBC 2013)	Radioactive water

Many of these incidents could be described with the same report, changing only the place, date, and chemical name.

In their investigation of the Bayamón incident (Figure 4.2), the CSB found poor maintenance of level indicators and alarms, inadequate redundancy, and a poor safety management system, citing similarities to the Buncefield incident 4 years earlier. The facility failed to maintain their level indicators, relying instead on manual calculations to estimate level. The facility also failed to maintain the secondary containment barriers, leaving the dike drain valve open, allowing spilled gasoline to enter the waste treatment plant, where the vapors ignited. Additionally, the CSB noted that a safeguard protecting against hurricane-force winds may have exacerbated the consequences of the release (CSB 2015):

Figure 4.2 Bayamón Fire (Source: CSB 2015)

> Similar to the Buncefield incident, during the overflow, gasoline sprayed from the tank vents, hitting the tank wind girder and aerosolized, forming a vapor cloud, which eventually ignited.

Human error can also defeat barrier integrity. Such was the case in Illiopolis, IL, USA, in 2004, when an operator overrode a critical valve safety interlock on a pressurized vessel making polyvinyl chloride (CSB 2007a). The vinyl chloride liquid and vapor discharged into the plant and the vapor ignited, causing an explosion that resulted in five fatalities. Another three workers were injured.

This incident illustrates the importance of communication systems. According to the CSB report, the operator had been washing out a reactor on the upper level of the reactor building, then went to the lower level to open the drain valve and reactor bottom valve and remove the cleaning liquid. CSB concluded that the operator most likely made an error on the way to the lower level, turning in the wrong direction and ending up at a different set of reactors.

There, the operator was able to open the drain valve, but not the reactor bottom valve, which was fitted with a safety interlock to prevent it from opening when the reactor was pressurized. However, the worker, not realizing he was at the wrong reactor, attached a hose labeled Emergency Air that provided the pressure needed for an override. That resulted in the release of the vinyl chloride.

Overriding the interlock was supposed to be done only in the event of an emergency. Furthermore, there were no gauges, indicators, or warning lights to inform operators on the lower level of each reactor's operating status. Communication between the operators on the lower level and upper-level-was nearly non-existent.

With the release of the highly flammable vinyl chloride, detection alarms sounded in the area. However, the supervisor and operators had not been trained to immediately evacuate the area in the event of a release. Instead, they stayed and unsuccessfully attempted to mitigate the release. As a result, they became casualties.

Situations do frequently arise where operator actions are a critical barrier, and sometimes the only barrier standing in the way of a catastrophic incident. Humans can serve as critical barriers, but only if they can be reliably notified of the need to respond; have the time to respond; know how to respond; can do it without putting themselves in danger; and are not under baseline stress (CCPS 2014).

See Appendix index entry S2

In Jaipur, Rajasthan, India, in 2009, two separate human barrier failures led to 11 people dead and more than 150 injured, many offsite. An operator was tasked to change the position of a spectacle blind to direct the transfer of gasoline to a desired location. The operator failed to block in the blind using two valves provided for this purpose. Upon swinging the blind, gasoline spewed into the dike. Unfortunately, the dike drain valve had been left open after a manual draining of rainwater. Gasoline vapors ignited and exploded and, unconfined to the dike, the fire spread widely.

The probability of human barrier failure increases with fatigue and stress. Sleep deprivation and distraction due to personal issues can lower a worker's decision-making ability and reduce reaction time. These situations are hard to guard against, so facility design should include one or more automated barriers in case human reactions are not correct or fast enough. Ensuring that such barriers are in operational order is paramount.

See Appendix index entry C46

This was apparent in the case of the 2010 Macondo oil well incident (CSB 2016a). Located 50 miles off the coast of Louisiana in the Gulf of Mexico, the oil well had multiple barriers that were supposed to prevent uncontrolled release of oil and gas. At the time, the crew onboard the rig was working on temporary well-abandonment activities. However, several barriers failed:

- A cement plug that had been put in earlier to keep the oil and gas below the sea floor was not installed effectively due to time pressure. The data from integrity testing of this barrier was misinterpreted. The cement barrier was incorrectly deemed effective.
- When the drilling mud was removed, hydrocarbons flowed past the blowout preventer (BOP) valve. There was no human detection (and thus no intervention) for nearly an hour. Once the crew did detect the release, they manually closed the BOP. By that time, however, the hydrocarbons had reached the surface and found an ignition source.
- An automated emergency response system worked as designed to shear the drill pipe and effectively seal the well. Pressure conditions in the well caused the drill pipe to buckle, however, rendering this barrier ineffective.

The failure of all these barriers resulted in 11 fatalities, 17 injuries, approximately 4 million barrels of hydrocarbons released into the environment, and an economic impact on the company of at least USD 65 billion (Maritime Executive 2018).

4.4 Chemical Reactivity Hazards

A chemical reactivity hazard is defined as a situation with the potential for an uncontrolled chemical reaction that can result directly or indirectly in serious harm to people, property, or the environment. The uncontrolled chemical reaction may be accompanied by a temperature increase, pressure increase, gas evolution, or other form of energy release. If heat is not properly removed from a process, it can result in a runaway reaction. Catastrophic incidents such as Seveso, Italy in 1976, Bhopal, MP, India in 1984 and Lodi, NJ, USA in 1995 were the result of runaway reactions.

See Appendix index entry C42

In October 2002, the CSB issued a report focusing on improving reactive hazard management using data collected from 1980 to 2001 (CSB 2002). The CSB data showed an average of six injury-related incidents per year in the USA, resulting in an average of five fatalities per year. More than 40% of these incidents were attributed to inadequate hazard evaluation resulting in a lack of awareness.

Following the CSB report, CCPS published *Essential Practices for Managing Chemical Reactivity Hazards* (CCPS 2003). CCPS also offers a free chemical reactivity worksheet (CCPS 2019) to help process safety managers identify reactive hazards. Even with all the chemical reactivity incident case histories and other tools available to help personnel better understand reactivity hazards, these incidents continue to occur. Greater awareness of chemical reactivity hazards and greater understanding of reactivity hazard scenarios is still critically important.

The reactivity of ammonium nitrate continues to be the source of major fatal incidents. Major accidents date back to 1923 in Oppau, Germany, where about 500 people died and thousands more were injured (BASF 2020). Twenty-four years later in Texas City, TX, USA, a fire started on a ship carrying 2,300 tons of ammonium nitrate (KTRK 2020); the result was like Oppau, with hundreds of fatalities and thousands injured. In 2001 at a plant in Toulouse, France, there was another major incident involving ammonium nitrate that, again, injured thousands of people while causing more than 30 fatalities (Guichon 2012). The 2013 West, TX, USA, incident also involved ammonium nitrate (CSB 2016b). As recently as the winter of 2019, a truck containing ammonium nitrate exploded after the tires caught fire in Camden, AR, USA, resulting in the driver's death (Rddad 2019). And then finally, as this book was in final production, 2,750 metric tons of ammonium nitrate in a warehouse in the port of Beirut, Lebanon exploded, killing at least 135 people, injuring at least 5,000, and causing damage in a 10-kilometer radius (Noack 2020).

See Appendix index entry C61

But ammonium nitrate far from the only chemical reactivity hazard. For example, the 2006 Morganton, NC, USA runaway reaction incident was the result of scaling up polymer production without properly accounting for the heat released from the reaction (CSB 2007b). The CSB investigation concluded that not only was there a lack of hazard recognition, but also poorly documented process safety information and ineffective control of product recipe changes. The report noted that the company did not conduct a process hazard analysis (PHA) and historically scaled up by simple trial and error without fully documenting the process.

See Appendix index entry C62

Similarly, in 2007, a company in Jacksonville, FL, USA, did not perform a PHA during scale-up of its methylcyclopentadienyl manganese tricarbonyl production (CSB 2009b). The company should have been aware of the potential for runaway reactions; several unexpected exothermic reactions had occurred prior to the start of commercial operations. However, rather than isolate the problem and redesign the process, the company attempted to control these exothermic reactions as they occurred. Unfortunately, this approach resulted in four fatalities and 32 injuries.

4.5 Asphyxiation in Confined Spaces

Inert gases such as nitrogen are commonly used for blanketing purposes to eliminate the possibility of fires or explosions and decrease evaporation. However, if the concentration of nitrogen is too high in the air we breathe, it can cause oxygen deprivation, asphyxiation, and toxic effects.

See Appendix index entry C68

A deadly incident involving nitrogen occurred in Hahnville, LA, in 1998. Workers were inside a temporary enclosure that had been built over the end of large gas pipeline (CSB 1999). They were not aware that the pipe was being purged with nitrogen, creating an oxygen-deficient atmosphere. One worker died and the other was seriously injured when they were asphyxiated. A CSB safety bulletin (CSB 2003) issued after this incident reported that, of the 85 workplace nitrogen asphyxiation incidents that occurred between 1992 and 2002, 62% occurred in the process industries.

Many asphyxiation incidents are caused by inadequate knowledge of the hazard when nitrogen is inadvertently used instead of an air delivery system. Such a case occurred in a large petrochemical plant in India. Workers intended to purge a distillation column with air before entering the column to do maintenance work. However, the air hose was accidentally connected to a nitrogen connection. The two fitters who entered the column died of asphyxiation (Venkataram 2020).

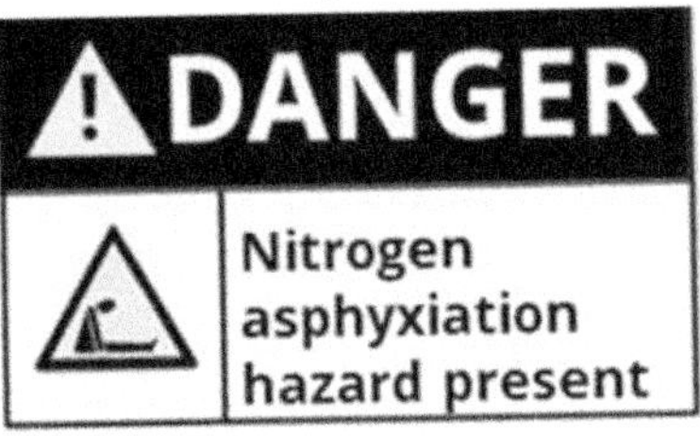

Figure 3.3 Signs Can Save Lives

To increase awareness of nitrogen hazards, many countries have regulations requiring control of confined spaces that mandate employers to identify all confined spaces in their workplace. Such regulations are specific, requiring warning signs such as the one shown in Figure 3.3 to be posted in permit-required confined spaces. They also require testing or monitoring to ensure acceptable entry conditions.

See Appendix index entry C71

Despite such national regulations and publicly issued incident reports, nitrogen asphyxiation incidents still occur. In the 2005 refinery incident in Delaware City, DE, USA, two contractors who were preparing to re-install a pipe on a pressure vessel while it was being purged with nitrogen died. According to the CSB findings (CSB 2006), the contractors knew that they needed a confined space entry permit, and that obtaining the permit would likely delay their work. Rather than go through the permit process, they tried a workaround solution that involved leaning into the reactor. One contractor either fell in or intentionally entered the reactor and was overcome by the lack of oxygen. The second contractor, seeing the other contractor in the reactor, tried to retrieve him. Unfortunately, he was also overcome, and both died.

4.6 Hot Work Hazards

The National Fire Protection Association (NFPA) defines hot work as work involving burning, welding, or a similar operation that is capable of initiating fires or explosions. Hot work is included in CCPS's RBPS Safe Work Practices element. Regulations and consensus standards for hot work are provided in the USA by US OSHA (29 CFR 1910.252), NFPA (NFPA 51B), and the American Welding Society (ANSI Z49.1), with similar regulations around the world. These documents establish best practices for effective hot work safety programs. But more needs to be done to protect workers during hot work activities.

In February 2010, the CSB issued a Safety Bulletin (CSB 2010) on preventing deaths during hot work in and around tanks. The bulletin was issued after the CSB saw that there was an upward trend in incidents, with seven accidents occurring between 2008 to 2010 alone. The CSB bulletin covered 11 incidents in various industries including refining, oil and gas, paper, food production, and wastewater treatment. It also outlined seven key lessons that could be learned from these incidents:

1. Use alternatives.
2. Analyze the hazards.
3. Monitor the atmosphere.
4. Test the area.
5. Use written permits.
6. Train thoroughly.
7. Supervise contractors.

Despite the abundance of information available, hot work incidents continue to occur, like the one in 2017 in DeRidder, LA, USA, where the explosion of a foul condensate tank caused the deaths of three contractors and injured seven others. The CSB report indicated that the tank likely contained water with a layer of flammable liquid turpentine floating on top and an explosive vapor space containing both air and flammable turpentine vapor above (CSB 2018). On the day of the incident, contractors were making repairs by welding on water piping above the tank, and this hot work was the probable source of ignition. In this case a hot work permit had been issued to the contractors; what was missing was a full job safety analysis that included the contents of the tank.

As this book is being written, there are two ongoing CSB hot work investigations (CSB 2020). One involves an explosion in May 2018 at a specialty-chemicals manufacturing plant in Pasadena, TX, USA, that injured 21 workers. The other is a flash fire that occurred in August 2016 at a terminal facility in Nederland, TX, USA, injuring seven workers.

4.7 References

4.1 API Recommended Practice (RP) 753 (2007). *Management of Hazards Associated with Location of Process Plant Portable Building*. Washington, DC: American Petroleum Institute.

4.2 API Recommended Practice (RP) 752 (2009). *Management of Hazards Associated with Location of Process Plant Permanent Buildings*. Washington, DC: American Petroleum Institute.

4.3 API Recommended Practice (RP) 756 (2014). *Management of Hazards Associated with Location of Process Plant Tents*. Washington, DC: American Petroleum Institute.

4.4 ARIA (2013). Overflow of a gasoline tank inside a refinery. IMPEL - French Ministry for Sustainable Development, Report No. 41148.

4.5 ASF (2000). Explosion caused due to an overflow of aqueous hydrogen peroxide at a peracetic acid manufacturing plant. www.shippai.org/fkd/en/cfen/CC1000131.html (accessed April 2020). (See Appendix, index entry J85).

4.6 BASF (2020) Explosion in Oppau. www.basf.com/ global/en/ who-we-are/history/chronology/1902-1924/1921.html (accessed May 2020).

4.7 BBC (2013). Fukushima leaks: radioactive water overflows tank. www.bbc.com/news/world-asia-24377520 (accessed April 2020).

4.8 Johnson, R.W., Rudy, S.W., and Unwin, S.D. (2003). *Essential Practices for Managing Chemical Reactivity Hazards*. New York: CCPS.

4.9 CCPS (2014). *Guidelines for Initiating Events and Independent Protection Layers in Layer of Protection Analysis*. Hoboken, NJ: AIChE/Wiley.

4.10 CCPS (2018). *Essential Practices for Creating, Strengthening, and Sustaining Process Safety Culture*. Hoboken, NJ: AIChE/Wiley.

4.11 CCPS (2020). Chemical Reactivity Worksheet. www.aiche.org/ccps/resources/chemical-reactivity-worksheet (accessed December 2019).

4.12 CSB (1999). Union Carbide Corp. Nitrogen Asphyxiation Incident. CSB Report No. 98-05-I-LA.

4.13 CSB (2002). Improving Reactive Hazard Management. CSB Report No. 2001-01-H.

4.14 CSB (2003). Hazards of Nitrogen Asphyxiation. CSB Report No. 2003-10-B.

4.15 CSB (2006). Valero Refinery Asphyxiation Incident. CSB Report No. 2006-01-I-DE.

4.16 CSB (2007a). Formosa Plastics Vinyl Chloride Explosion. CSB Report No. 20014-10-I-IL.

4.17 CSB (2007b). Synthron Chemical Explosion. CSB Report No. 2006-04-I-NC.

4.18 CSB (2009a). INDSPEC Chemical Corporation Oleum Release. CSB Report No. 2009-01-I-PA.

4.19 CSB (2009b). T2 Laboratories Inc. Reactive Chemical Explosion. CSB Report No. 2008-3-I-FL.

4.20 CSB (2010). Seven Key Lessons to Prevent Worker Deaths During Hot Work in and Around Tanks. CSB Report No. 2009-01-SB.

4.21 CSB (2015). Caribbean Petroleum Tank Terminal Explosion and Multiple Tank Fires. CSB Report No. 2010.02.1.PR.

4.22 CSB (2016a). Macondo Blowout and Explosion. CSB Report No. 2010-10-I-OS.

4.23 CSB (2016b). West Fertilizer Explosion and Fire. CSB Report No. 2013-02-I-TX.

4.24 CSB (2018). Explosion at PCA's DeRidder, Louisiana, Pulp and Paper Mill, 2018. CSB Report 2017-03-I-LA.

4.25 CSB (2019a). Toxic Chemical Release at the DuPont LaPorte Chemical Facility. CSB Report No. 2015-01-I-TX.

4.26 CSB (2019b). Storage Tank Fire at Intercontinental Terminals Company, LLC (ITC) Terminal: Factual Update. CSB Report No. 2019-01-I-TX.

4.27 CSB (2020). Current investigations. www.csb.gov/investigations/current-investigations (accessed May 2020).

4.28 Dole, E. and Scannell, G.F. (1990). *Phillips 66 Company Houston Chemical Complex Explosion and Fire: Implications for Safety and Health in the Petrochemical Industry: A Report to the President*. Washington, DC: US Department of Labor.

4.29 Doyle, A. (2020). Hundreds hospitalised after styrene gas leak in India. www.thechemicalengineer.com/news/hundreds-hospitalised-after-styrene-gas-leak-in-india (Accessed May 2020).

4.30 Guichon, G. and Jacob, L. (2012) On the catastrophic explosion of the AZF plant in Toulouse (21 September 2001). *Proceedings of the Global Congress on Process Safety,* Houston, Texas (1– 5 April 2012). New York: AIChE.

4.31 Health and Safety Executive (2011). Buncefield: Why did it happen? COMAH Competent Authority. www.hse.gov.uk/comah/buncefield/buncefield-report.pdf (Accessed April 2020).

4.32 Health and Safety Executive (2020). PEMEX LPG Terminal, Mexico City, Mexico. 19th November 1984. COMAH Competent Authority. www.hse.gov.uk/comah/sragtech/casepemex84.htm. (Accessed April 2020).

4.33 KTRK (2020). Remember when: Hundreds killed, thousands injured in Texas City disaster of 1947. abc13.com/disaster-fire-explosion-texas-city/1865491. (Accessed May 2020).

4.34 Laris, M., Duncan, I. and Aratani, L. (2019). FAA Administrator Dickson pressed on agency's prediction of 'as many as 15' Max crashes possible. *Washington Post* (11 December 2019).

4.35 Maritime Executive (2018). BP's Deepwater Horizon costs reach $65 billion. www.maritime-executive.com/article/bp-s-deepwater-horizon-costs-reach-65-billion. (Accessed April 2020).

4.36 Noack, R. and Mellen, R (2020). What We Know About the Beirut Explosions. www.washingtonpost.com/world/2020/08/05/faq-what-we-know-beirut-explosions. (Accessed August 2020).

4.37 Quintão, A. (2019). CPI da Barragem de Brumadinho. www2.camara.leg.br/atividade-legislativa/comissoes/comissoes-temporarias/parlamentar-de-inquerito/56a-legislatura/cpi-rompimento-da-barragem-de-brumadinho/documentos/outros-documentos/relatorio-final-cpi-assembleia-legislativa-mg. (Accessed August 2020).

4. 38 Rddad, Y. and Snyder, J. (2019). Driver killed after fertilizer truck explodes in South Arkansas; Area evacuated after blast that was heard miles away. *Arkansas Democrat Gazette* (27 March 2019).

4. 39 Robertson, P.K. de Melo, L., Williams, D. et al. (2019a). Report of the Expert Panel on the Technical Causes of the Failure of Feijão Dam I. Agência Nacional de Mineração.

4. 40 Robertson, P.K. (2019b). Transcript of Video presentation by Dr. Peter K. Robertson, Ph. D., Chairperson of the Expert Panel on the Technical Causes of the Failure of Feijão Dam I. Agência Nacional de Mineração.

4.41 United Kingdom Department of the Environment (1975). *The Flixborough Disaster*. London, UK: Her Majesty's Stationary Office.

4.42 Venkataraman, K. (2020). Gas leak at Haldia Plant, 3 workers die. Indianexpress.com/article/cities/kolkata/gas-leak-at-haldia-plant-3-workers-die/ (Accessed June 2020).

5
LEARNING MODELS

"A donkey that carries a lot of books is not necessarily learned."
—Danish Proverb

How does a company study and learn? It doesn't. Individuals study and learn. Individual employees can teach their coworkers and spread what they have learned. But companies can only be said to have learned when they make a permanent change in the way they operate. This corporate change is the goal of evaluating internal and external incidents.

Many individual learning models exist, as do many corporate change models. We discuss a representative sampling of each in this chapter. However, ultimately, individual learning and corporate change together are not tied together in any model found while developing this book. We therefore sought to adapt the good features and practices in existing models to a new model that meets this book's objective. In this chapter, we discuss the features and practices of a range of models, extract the useful ones, and assemble these together to form a new model for driving continuous process safety improvement from investigated incidents.

5.1 Learning Model Requirements

An effective learning model to help drive continuous process safety from investigated incidents should do the following:

- *Empower* leaders, frontline personnel, and process safety professionals to study past incidents, both external and internal.
- *Focus* on areas of improvement that can have a significant impact on corporate process safety performance.
- *Direct* lessons learned into appropriate corporate systems, e.g. the PSMS, standards, and policies.
- *Motivate* adherence to these corporate systems over time.
- *Address* the obstacles to learning discussed in Chapter 3.

No individual or company can be expected to learn from all 441 publicly available incident reports identified in this book, the countless others potentially available, and all the company's internal incident and near-miss reports. Therefore, the model should provide a mechanism to allocate learning resources and select incidents with learnings directed towards specific, meaningful corporate improvement goals. This will help obtain the largest impact for the effort and reinforce that the effort adds value.

As discussed in Section 3.2, one reason companies fail to learn is that the knowledge gained by experts stays in the ivory tower. An effective learning model must ensure that findings and recommendations developed by experts are translated into corporate systems, including the PSMS, corporate standards, and corporate policies. This can only happen by integrating individual learning with overall corporate change.

Driving adherence to the PSMS, standards, and policies is a leadership responsibility and a responsibility of all employees. However, because process safety incidents occur less frequently than occupational safety incidents, process safety tasks, training, and reminders may be treated as less important, causing institutional knowledge to fade. An effective learning model must provide for continuous motivation and reminders to keep knowledge fresh.

The process of corporate learning from incidents should both align individual work to corporate goals and drive corporate actions and individual work. In industry, the well-known framework for achieving this is the quality improvement cycle of Plan-Do-Check-Act (PDCA).

It is worth noting that the work in the PDCA cycle can be divided between actions by individuals and action by the company. The work can be further divided between gathering and analyzing facts and information versus acting. This division of work is depicted in Figure 5.1.

	Individual	Company
Gather facts	Study **II. Do**	Focus—Objectives **I. Plan**
Interpret and act	Develop and recommend **III. Check**	Implement **IV. Act**

Figure 5.1 Mapping Desired Learning Framework to Plan-Do-Check-Act

Sections 5.2 and 5.3 will discuss a representative sampling of individual learning models and corporate change models, highlighting the most useful features to include in a new model for driving corporate learning from investigated incidents. Specifically, Section 5.2 will describe learning models, which tend to address work done by individuals (boxes II and III in Figure 5.2); Section 5.3 will describe corporate change models, which tend to address actions taken by the company (boxes I and IV).

5.2 Learning Models for Individuals

This section presents a representative selection of learning models for individuals and small groups and identifies the characteristics of each model that are useful for driving continuous process safety improvement.

5.2.1 Multiple Intelligences and Learning Styles Model

Research by the American psychologist Howard Gardner led him to identify of eight independent forms of intelligence (Gardner 1995 and 2011) summarized in Table 5.1. Each person has one or more of these intelligences, but generally not all of them. Gardner showed that everyone learns best when learning engages his or her specific forms of intelligence.

Table 5.1 Gardner's Eight Forms of Intelligence

Intelligence Type	Summary
Musical	Attuned to tones, rhythms, and harmonies, often has perfect pitch
Visual-spatial	Able to visualize in the mind
Verbal-linguistic	Attuned to reading, writing, telling stories; good at memorizing
Logical-mathematical	Good at abstraction, logic, and numbers. Connects cause and effect
Kinesthetic	Control of physical motion, manual dexterity
Interpersonal	Perceptive to other's feelings, prefers to work in groups
Intrapersonal	Reflective, prefers to work alone
Naturalistic	Recognizes how biological and physical systems work together

Gardner's work has led many other educators to develop evaluation and teaching tools based on this framework. Typically categorized as following the Learning Styles Model, these tools help educators and trainers present lessons and communications in forms that best fit their students' intelligences, in order to boost learning.

Gardner's theory of multiple intelligences was intended to be used primarily in box III (Check) of Figure 5.1. However, since learning and communication are opposite sides of the same coin, Gardner's model should also be extremely useful for helping to refresh learning, a key feature needed in box IV of the model (Act).

In all but the smallest companies, we can expect that all eight forms of intelligence—and therefore all eight learning styles—will exist among our leadership and workforce. It follows that to effectively drive learning about process safety across the company, we need to communicate using multiple styles. This will also help promote workforce involvement.

5.2.2 Career Architect® Model

Lombardo and Eichinger have developed a model for individual learning and development in a corporate setting (Lombardo 2010) in which individuals partner with their managers (and possibly with mentors) to identify competencies they should develop and career hurdles they can overcome.

The learning-and-development plan the employee and manager agree on focuses on providing the employee with on-the-job experiences in the areas where development is needed. During these experiences, about 70% of an employee's learning comes from doing, 20% by interacting with and observing those around them, and only 10% from courses and reading.

The model identifies 167 different competencies an employee may have to develop. For each, Lombardo describes a process to assess the need for development, then suggests means of learning from self-study, feedback, development in the current position, and development in a new position.

The Career Architect® Model's focus on a broad list of possible employee and company objectives links corporate goals in box I (Plan) and box II (Do) of Figure 5.1. The 167 potential development competencies represent an important use of metrics to guide where learning should take place. Use of metrics is a necessary feature of the new learning model that is the centerpiece of this book.

5.2.3. Dynamic Learning

In the digital age, classroom-based training and online slide decks are a thing of the past. New cloud-based digital learning experiences are flexible, personalizable, effective, and good for the bottom line (Willyard 2016).

As with the Career Architect model, employees and their managers work together to identify learning and development needs. Then, development is supplemented with dynamic novel online tools, including:

- *micro-learning*. Short lessons, typically three minutes or less, reflect the trend in today's society toward consuming information as short videos or social media posts. Micro-learning is best when applied to very concise learning opportunities, for example to tackle an unfamiliar task in a new assignment, or as a periodic refresher. It may also be used to segment longer courses into manageable bites.
- *self-serve learning*. E-learning is available on demand when the employee needs it.
- *learning as entertainment*. Digital games, virtual reality, interactive presentations, and competitions are effective learning tools—and have special appeal for younger workers.
- *social learning*. Online collaboration platforms make it easy for employees to share and collaborate.
- *user-generated content*. Content doesn't have to be polished to be effective. Students today are eager to develop their own content—presentations, videos, signs, and more—and to learn from other students.

Like Gardner's Multiple Intelligences Model, the Dynamic Learning Model fits primarily in box III (Check) of Figure 5.1 but should be extremely useful in box IV (Act). Dynamic Learning serves as a reminder that our toolbox for communicating learning experiences is larger than the classroom—and likely to keep expanding. When we review internal and external incidents, we should look beyond reports and videos, and when institutionalizing the learning, we should not limit the way we communicate.

5.2.4 Ancient Sanskrit

An ancient Sanskrit *shloka* states (Surya Rao 2019; Joshi 2009):

आचार्यात् पादमादत्ते पाद शिष्य स्वमेधया
पाद सब्रह्मचारिभ्य. पाद काल क्रमेण च

One gains ¼ of the knowledge from the Acharya (the teacher), ¼ from his own self-study and intellect, ¼ from co-learners and the balance out of experience.

This ancient bit of wisdom recognizes the distribution of responsibilities for learning between teachers and their students. It also recognizes an important difference between education (the process of gaining knowledge) and training (turning that knowledge into practical skills).

In the context of learning from incidents, the teacher is the organization that publishes the incident report and/or video. The teaching comes when the organization highlights the key findings of the investigation and translates these findings into recommendations. The student is the process safety expert working for the company who reads the report and studies the video. A co-dependent relationship therefore exists between teacher and student. Without the teacher, the student cannot access important information. But if the students don't do their part, the teaching goes into a vacuum.

The fellow students in this model correspond to the other subject matter experts or discipline leaders with whom the process safety expert shares the report and discusses its implications. Together, they consider the findings and recommendations and study deeper, in particular to determine any learnings from the incident that might not have been emphasized by the investigative organization yet might be valuable to the company.

In the Ancient Sanskrit model, experience comes from the way the company runs its operations. The company expresses its experience in multiple ways, including the PSMS, standards, and policies, but also in procedures, designs, the complete base of process safety knowledge, and, of course, the culture. The above-mentioned team of subject matter experts recommends appropriate changes in these areas.

The Ancient Sanskrit model primarily addresses boxes II and III of Figure 5.1 (Do and Check). It also informs and is informed by boxes I and IV (Plan and Act).

5.2.5 Guiding Principles for Learning

Repsol, a global energy company based in Madrid, Spain, provides its leaders and employees with the following guiding principles for growing the organization's safety competence (Blardony-Arranz 2020):

- lead
- capture
- share
- learn.

In this model, leaders facilitate the learning process, actively driving individual learning toward corporate improvement. Leaders support employee learning by providing time and learning resources, promoting an environment that encourages questioning, and showing visible support. Leaders also facilitate change in response to learning.

The model treats any learning brought to the organization as a gift that must be captured. Employees are encouraged to report hazards and weak barriers, to seek opportunities to improve human factors, and to identify inherently safer options.

Sharing is a key dimension of the model. This emphasis not only facilitates communication among team members but also helps drive better solutions by involving people with diverse perspectives and skill sets.

The model considers the process of learning to be one of sustainable action. Learning occurs only when the process, operations, and culture change so that the improvement will persist over time.

Repsol's Guiding Principles touch all four quadrants of Figure 5.1. The model's over-arching theme is that learning must be intentional, driven by corporate needs, supported by open communication, and permanently maintained by intentional action.

5.3 Corporate Change Models

This section presents a representative selection of corporate change models that apply to a variety of changes and corporate cultures. As in the previous section, we identify the characteristics of each model that apply to the combined model we seek.

5.3.1 Lewin

Kurt Lewin's 1948 pioneering work described organizational change in three simple stages (Levasseur 2001):

- unfreeze
- change
- refreeze.

The unfreezing step includes recognizing the need for improvement. This is best accomplished by understanding what will make the people across the company want to change. The process includes obtaining data about where improvement is needed and, if possible, where the greatest leverage can be obtained.

The change step involves determining how to change the organization. This, too, is a participative step, involving everyone in the improvement efforts.

The refreeze step addresses the follow-through needed to implement and then maintain the change by anchoring it in the culture. This can be the most difficult step, because it requires that people get used to doing things a new way—while old habits die hard. Leadership must provide ongoing support, model the new way of doing things, and monitor for signs of regression.

Relative to the desired learning model, Lewin's Unfreeze and Freeze steps map to boxes I and IV (Plan and Act) of Figure 5.1. In other words, the Lewin model describes well what needs to change and what the result should be, but it leaves out the details.

5.3.2 McKinzie 7-S®

The McKinzie 7-S®model, named for the management consulting firm that developed it and first described in Tom Peter's landmark book *In Search of Excellence* (Peters 2004), considers corporate change at a more anatomical level. The McKinzie® model looks for weaknesses in seven dimensions of corporate performance: strategy; structure; systems; shared values; style; staff; and skills.

The model also considers weaknesses in the interfaces between each of the dimensions. Related to the desired learning model, McKinzie® tends to address most strongly box I (Plan) of Figure 5.1. As weaknesses are identified, leaders implement incremental changes over time. By making smaller, slower changes, this model avoids requiring major culture shifts and is therefore unlikely to meet resistance or reversion to past practices.

The McKinzie® model may not be ideal for driving change from process safety findings and recommendations. Because process safety incidents are already relatively rare compared to other business events, McKinzie® could contribute to the sense that improvement is not urgently needed.

Nonetheless, McKinzie® considers corporate dimensions that are often not factored into the PSMS, standards, and policies, most notably strategy, structure, and style. When developing goals for the learning process, it could prove useful to seek opportunities in these areas, not just the more traditional ones we consider when improving process safety.

5.3.3 Kotter

John P. Kotter, the Harvard professor emeritus who is sometimes described as the world's foremost authority on leadership and change, developed a model that focuses on how leaders drive change at a high level rather than getting into the details of what to change (Kotter 2012). Kotter guides leaders to:

- create a sense of urgency
- build a core coalition
- form a strategic vision
- get everyone on board
- remove obstacles and reduce friction
- generate short-term wins
- sustain acceleration
- set the changes in stone.

Related to the desired learning model, Kotter's model tends to address boxes I (Plan) and IV (Act) of Figure 5.1.

Clearly, Kotter's model applies for companies and plants that need a major overhaul of their PSMS, standards, policies, and even organizational structure. Sometimes, however, companies need only minor changes, for which Kotter's model would be overkill. Either way, the learning process we seek must be part of a broader top-down process safety effort. If such an effort has not yet been implemented, the Kotter model would be a highly effective way to deploy it.

5.3.4 ADKAR®

The ADKAR® model developed by the change management company Prosci is designed to drive change from the bottom up (Hiatt 2006). The ADKAR® acronym stands for the first letters of:

- awareness of the need to change
- desire to participate and support the change
- knowledge on how to change
- ability to implement required skills and behaviors
- reinforcement to sustain the change.

Related to the desired learning model, ADKAR® tends to address boxes III and IV (Check and Act) of Figure 5.1.

Most process safety leaders today are accustomed to driving change from the bottom up and should feel familiar with ADKAR®. However, CCPS recommends that process safety be driven from the top of the company through, middle managers, to the frontline (CCPS 2019).

5.3.5 IOGP

A working group of the International Association of Oil and Gas Producers (IOGP) has developed a guide based on how their member companies typically learn from process safety incidents (IOGP 2016). Although this guide focuses on learning from internal process safety incidents, the learning processes it identifies describe the current state of how learning occurs in the international oil and gas sector learn.

The IOGP working group identified a set of principles that guide how companies should think about learning from incidents. (They also developed a map of components, that is, an inventory of how member companies learn.) The 10 guiding principles include:

- *Something must change*. If the company plans to learn from incidents, it must commit to changing what it does based on what's learned.
- *Learning is one way we manage risk*. Learning allows companies to control risks better.
- *Sharing is not the same thing as learning*. Just because you pass on information, that does not mean the recipient changes behavior.
- *Balance short-term temporary mitigation with long-term sustainable response*. Because long-term solutions take time, short-term interim actions may be needed while the long-term solutions are developed.
- *It's necessary to focus*. Considering the wide range of potential learning, the company needs to focus on what can really make a difference.
- *Learning is a distributed effort*. No one person or group can drive all learning for a company.
- *Don't restrict learning to individual incidents*. Look for patterns and themes across multiple incidents.
- *Promote collaboration*. Developing solutions is as much a distributed effort as is learning.
- *Leaders make learning work*. People do what their leaders value. For learning to work, leaders must value learning.
- *Close the loop*. Manage the learning process. Ensure efforts are paying off as expected and adapt as needed.

The IOGP map of components describes four steps in the learning process:

- recognize and report events
- investigate events
- systematically apply learning
- support self-motivated learning.

The IOGP model may reflect the tendency among companies in the petroleum producing sector to use a high level of common technologies and equipment across their operations. This standardization allows outside technology providers to do the high-level strategy and R&D, freeing the companies to focus their activities in boxes III and IV (Check and Act) of Figure 5.1.

5.4 The Recalling Experiences and Applied Learning (REAL) Model

As evident from the previous two sections, no one learning or corporate change model meets this book's objective of describing a novel, comprehensive highly effective learning model for driving continuous process safety improvement from investigated incidents, as set out in Section 5.1. However, all the pieces are there to create exactly what is needed.

Figure 5.2 summarizes how each of the above-described models can contribute to a comprehensive new model for driving continuous process safety improvement from investigated incidents–both internal and external.

	Individual	Company
Gather facts	**Study** Ancient Sanskrit *Guided Principles* *IOGP* **II. Do**	**Focus—Objectives** Learning Styles, Lewin, Kotter, Career Architect®, McKinzie 7- S®, *Ancient Sanskrit, Guided Principles* **I. Plan**
Interpret and act	**Develop and Recommend** ADKAR®, IOGP, *Ancient Sanskrit, Guided Principles* **III. Check**	**Implement** Dynamic Learning, Lewin, Kotter, ADKAR®, IOGP, *Ancient Sanskrit, Guided Principles* **IV. Act**
Legend	Normal font = directly applies	*Italic = indirectly applies*

Figure 5.2 Potential Contributions of Learning and Change Models

Extracting the desired features from the models described in Sections 5.2 and 5.3 and applying the collective experience of the members of the CCPS subcommittee, we developed a new model, the Recalling Experiences and Applied Learning (REAL) Model. Figure 5.3 shows how the desirable features from the other models fit in the overall PDCA structure.

	Individual	Company
Gather facts	2. Seek learnings 3. Understand 4. Drilldown **II. Do**	1. Focus **I. Plan**
Interpret and act	5. Internalize 6. Prepare **III. Check**	8. Embed and Refresh 7. Implement **IV. Act**

Figure 5.3 Recalling Experiences and Applied Learning (REAL) Model

The REAL Model has eight components:

1. ***Focus.*** Company leadership commits to improving process safety performance, including implementing learnings from incidents. Based on metrics and metric trends, audit findings, workforce involvement, and other sources of input as well as self-assessment of corporate strategy, structure, and style (Peters 2004), leaders determine focused goals for improvement. These goals, which may be either incremental or fundamental, guide the learning process.
2. ***Seek learnings.*** One or more evaluators are provided with the necessary time and resources to evaluate incidents with the potential to provide learnings that address the corporate improvement goals.
3. ***Understand.*** Through individual study and, as needed, discussion with peers, the evaluator comes to deeply understand these incidents.
4. ***Drilldown.*** Because the evaluator may have different objectives than the authors of the incident reports, the evaluator seeks findings and recommendations that the authors did not elucidate in their report.
5. ***Internalize.*** The evaluator collaborates with company peers in appropriate functions to translate the key findings and recommendations into proposed changes to the company PSMS, standards, policies, and process safety knowledgebase.

6. ***Prepare.*** The evaluator works with peers to develop an action plan to implement the changes. The plan includes resources, capital, training, communication, and other key factors.
7. **Implement.** After company leadership has endorsed the plan, the plan is implemented. The implementation team may include both corporate experts and site leaders. Implementation includes leadership, workforce involvement, training, conduct of operations, metrics, and ongoing management review.
8. **Embed and Refresh.** Company and site leadership now manage according to the changes as implemented. Ongoing communications and training remind, maintain the sense of vulnerability, and reinforce the need to maintain the new way of doing things.

These components of the REAL Model will be described in greater detail in chapters 6 and 7. Chapters 9–14 will provide some hypothetical examples of how the REAL Model may be used.

5.5 References

5.1 CCPS (2019). *Process Safety Leadership from the Boardroom to the Frontline*. Hoboken, NJ: AIChE/Wiley.

5.2 Gardner, H. (1995). Reflections on multiple intelligences: myths and messages. *Phi Delta Kappan* 77: 200–209.

5.3 Gardner, H. (2011). *Frames of the Mind: The Theory of Multiple Intelligences*. New York: Perseus Books Group.

5.4 Hiatt, J.M. (2006). *ADKAR: A Model for Change in Business, Government, and Our Community*. Fort Collins, CO: Prosci Learning Center Publications.

5.5 International Association of Oil & Gas Producers (2016). Components of Organizational Learning from Events. IOGP Report No: 552.

5.6 Joshi, S. (2009). How we learn and grow. blog.practicalsanskrit.com/2009/12/how-we-learn-and-grow.html (accessed January 2020).

5.7 Kotter, J.P. (2012). *Leading Change*. Brighton, MA: Harvard Business Review Press.

5.8 Levasseur, R.E. (2001). People skills: Change management tools—Lewin's change model. *FOX Consulting Group Newsletter*, July-August 2001.

5.9 Lombardo, M. M. and Eichinger, R. W. (2010). *Career Architect Development Planner*, 5th Edition. Minneapolis, MN: Lominger.

5.10 Peters, T. and Waterman, Jr., R.H. (2004). *In Search of Excellence: Lessons from America's Best-Run Companies*. New York: Harper Business.

5.11 Willyerd, K., Grünwald, A, Brown, K. et al. (2016). A new model for corporate learning. *D!gitalist Magazine* (11 March 2016).

6
IMPLEMENTING THE REAL MODEL

"I read, I study, I examine, I listen, I reflect, and out of all this I try to form an idea into which I put as much common sense as I can."
—Marquis de Lafayette, French Nobleman and Military Officer

Successful execution of the REAL Model requires both individual evaluation and corporate change. The most basic requirements for implementation are:

- leadership support and involvement at all levels.
- enough people with the proper knowledge and experience evaluating external and internal incidents
- a workforce interested and motivated to improve process safety performance.

If you are reading this book, you are probably a member of the second or third group in the list. For you to have success in achieving lasting improvement for the company or plant, it will be important for you to obtain not only leadership support and involvement in setting objectives and driving the changes that the model identifies but also the needed financial and human resources.

If the company leadership team members have not yet bought into their roles in driving the PSMS, getting this support is an important first step. CCPS provides three helpful resources:

- *Process Safety Leadership from the Boardroom to the Frontline* (CCPS 2019a). This book lays out the business case for process safety, describes what leaders at each level must do to fulfill their roles, and helps dispel many misconceptions leaders may have.

- *Guidelines for Integrating Management Systems and Metrics to Improve Process Safety Performance* (CCPS 2016). This book dedicates an entire chapter to securing support for process safety improvement efforts.
- *Essential Practices for Building, Strengthening, and Sustaining Process Safety Culture* (CCPS 2018a). This book describes a step-by step process to understand the current process safety culture and drive improvement.

Getting buy-in for a program of learning from incidents should be easier once leaders take responsibility for driving the PSMS. Start by reminding your audience that with any incident, every preventive barrier and the associated management systems failed. With a near-miss, at least one preventive barrier and its associated management system failed, and the only reason the others did not fail may have been sheer luck. Digging into the details in the PSMS, standards, or policies to identify and correct the deficiencies that allowed these barriers to fail will prevent future incidents from occurring due to the same failures.

Fully involved leaders know that near-misses are "a gift...on a silver platter" (CCPS 2019), because learning from near-misses allows the company to drive improvements without bearing the high cost of an accident. But if learning from near-misses is a gift, learning from someone else's incident is an even bigger gift. After all, near-misses are seldom free: they often result in lost or reduced production, or the need to replace damaged equipment or instruments or rework an off-spec product.

Just to be clear, the REAL Model does not relieve the company of the need to evaluate internal incidents and near-misses or the need to remind personnel of and commemorate these incidents and their related institutional knowledge. Although the model focuses on driving improvement and developing preventive actions based on lessons learned from external incidents, it can and should also be applied to internal incidents.

Each of the steps of the REAL Model will be discussed in detail in Sections 6.1–6.8. Chapter 7 will then discuss in greater detail a range of ways to keep learnings fresh.

6.1 Focus

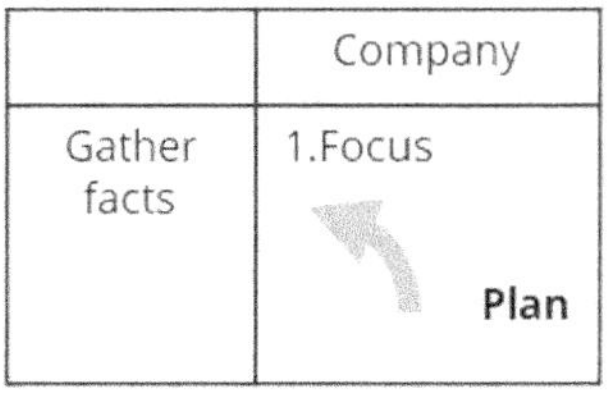

	Company
Gather facts	1.Focus **Plan**

The large and always-growing number of external incidents described in the media and the literature makes it impossible to know about every relevant past incident or stay current on new ones, let alone remember all the findings. Therefore, any corporate improvement effort based on learning from incidents should be focused on areas where the company can derive the most benefit. Identifying these areas is typically part of the process safety management system reviews (CCPS 2007) conducted by leadership teams. The next two subsections will discuss how companies can identify where to focus their learning and improvement activities.

6.1.1 Identifying High Potential Impact Learning Opportunities

High potential impact opportunities can be identified from:

- *landmark incidents.* Incidents like those discussed in Chapter 8 have timeless learning. The company should have implemented improvements from these incidents. It's important to review and verify that the knowledge remains institutionally strong.
- *leading metrics.* Tier 4 metrics collected to monitor the performance of the PSMS (API 2016; CCPS 2018b).
- *near-miss metrics.* Tier 3 metrics collected to monitor challenges to safety systems (Ibid.).
- *audit findings.* PSMS deficiencies found through periodic in-depth review (CCPS 2011a).
- *investigations.* Incident and near-miss investigations at the site and across the company (CCPS 2019b).
- *risks on the risk register.* The potential consequence and/or risk scenarios of greatest interest to the Board and senior leadership.
- *gut-feel.* Warning signs based on experience that should be acted-on.
- *high-profile incidents.* Other companies' incidents in similar plants and processes.
- *acquisitions.* Problems uncovered during due diligence evaluations and post-acquisition integration.
- *new products, processes, and designs.* New product or technology directions with unfamiliar challenges.
- *corporate change.* Anticipated downsizing, growth, or reorganization.

A company may, and probably does, have other drivers for seeking improvement. Let's examine each of the above opportunities in turn.

Landmark Incidents that Every Company Should Learn From

Several incidents, for example Bhopal and Piper Alpha, are so critical to our understanding of process safety that every company should work on an ongoing basis to ensure that PSMS, standards, policies, and culture embody their lessons learned. These incidents, which span the process industries and beyond, are described in Chapter 8.

Leading Metrics (Tier 4)

Ultimately, management systems describe tasks. Leading metrics, therefore, measure how faithfully, and how well, the company is performing its tasks. Assuming appropriate metrics have been defined (API 2016, and CCPS 2018b), any adverse Tier 4 metrics result should serve as a trigger to drive deeper investigation of why the PSMS is under-performing. Here are just a few examples (the references provide more):

Action item follow up

Action items routinely arise from PHA, MOC, incident investigations, and audits. Failure to follow up on action items indicates an area where improvement is needed. Some examples include:

- *Asset Integrity.* Inappropriate choice of inspection, testing, and preventive maintenance (ITPM) intervals, issues related to how ITPM is performed, and material of construction selection and/or verification.
- *Emergency Management.* Gaps in emergency training of personnel, coordination and training of off-site responders, inspection of emergency equipment and verification of adequate supplies (e.g., air bottles or hoses and equipment).
- *Hazard Identification, Consequence Analysis, and Risk Analysis.* Neglecting important hazard scenarios, for example, severe weather, not factoring in potential hazards and risks outside of the site, not fully understanding how the process was designed and how things could fail, overly optimistic estimation of barrier strength, independence, or failure frequency.

Near-Miss Metrics (Tier 3)

Tier 3 metrics track observable and usually measurable barrier failures and challenges to safety critical systems. When a barrier fails, it means there has been a failure of at least one, and possibly more than one, PSMS element, standard, or policy. Therefore, Tier 3 metrics clearly identify areas that require improvement.

Examples of Tier 3 metrics include:

- trips of Safety Instrumented Systems (SISs)
- activations of interlocks
- manual activation of emergency shut-down
- exceedance of safe operating limits
- any leak or release not included in Tier 1 or Tier 2.

Near-misses covered by Tier 3 metrics should be investigated as if an incident had occurred.

<u>Audit Findings</u>

Process safety audits for regulatory compliance are generally conducted every three years; for risk management purposes, audit frequency may range from one to five years, depending on risk. Audits conducted by leading companies typically go beyond compliance and may include (CCPS 2011a):

- more frequent audits of processes that have higher risk and/or higher potential consequences
- auditing processes not covered by regulations that nonetheless have appreciable risk
- benchmarking against industry-leading practices.

All audits capture deviations from the PSMS as findings; however, some will also capture recommendations for improving the PSMS, standards, and policies. Findings and recommendations can inform overall improvement efforts, especially those that recur through multiple sites. Although repeat findings often reflect deficiencies in how site management carries out process safety responsibilities, they may also reflect an inherent gap in site or corporate knowledge that could be informed by examining internal and external incidents.

<u>Internal Incident and Near-Miss Investigations</u>

Investigation of internal incidents and high potential near-misses should automatically trigger a review of both relevant external incidents and similar internal incidents. While this review could occur separately from the investigation, in many cases it can be beneficial to review other incidents during the investigation phase to aid in identifying root causes and causal factors; and then while developing recommendations.

The review of incidents should go beyond considering similar industries and processes to include those with similar root causes and causal factors, regardless of industry or process type.

Risk Register

Many companies maintain a risk register. This is a document or system in which process safety and other risks are determined (for process safety via PHA and Layer of Protection Analysis) and sorted. Leadership uses this information as a guide in managing operations and looking for improvement opportunities. It is natural to focus process safety learning on the highest process safety risks. However, many companies recognize the benefits of also addressing potential scenarios with the most severe consequences, regardless of risk. As process safety pioneer Trevor Kletz said on many occasions, "What you don't have, can't leak."

Gut Feel and Warning Sign Analysis

CCPS found that people interviewed during incident investigations frequently used the phrase, "I knew we were going to have that incident" (CCPS 2011b). Deeper inquiry showed that the interviewees came to that realization by noticing things that often don't show up in metrics, audits, or other ways of measuring process safety performance. During a conference commemorating the 25th anniversary of the North Sea Piper Alpha disaster, Steve Rae, an engineer who escaped the platform, talked about the warning signs he had seen (Rae 2013):

> I want to share with you my initial thoughts on my arrival on Piper Alpha and the three months after:
>
> 1. I looked at Piper Alpha on arrival and thought wow, it looks old and tired ... when in fact it wasn't old at all.
> 2. Obvious that additional structures and modules had been added.
> 3. Somewhat confusing to navigate around, somewhat of a rabbit's warren.
> 4. During nights there were few people around, in particular around the control room where I had to go permits to work.
> 5. There were times when Piper vibrated significantly, it almost felt like it would be shaken to pieces.

6. The airlock doors from the main deck to the office block were rarely closed.
7. Drillers very much left to their own devices. We had our own isolation procedures, lock out system, maintenance team.

Rae acknowledged that, although he didn't recognize it at the time, these were strong warning signs of breakdowns in management of change, emergency management, asset integrity, operational discipline, and safe work practices.

Why wait for an incident investigation to hear the warning signs that preceded the incident? Ask frontline and leadership personnel at the plant and site to imagine that they are being interviewed after a major incident. Then have them complete this sentence: "I knew that this incident would occur because _________."

You can also ask personnel, "What is the one thing that concerns you most about our present operations?" or "If you could improve one process safety task, what would it be?" These questions can help to identify potential incidents before they happen. Try this approach during task reviews, toolbox talks, and safety meetings, or in meetings specifically held for this purpose.

When employees at any level are engaged in these exercises, common concerns will emerge. They represent the collective gut feel of the plant or site—they are warning signs of where the PSMS may be weak. Any potential weaknesses that don't match up with metrics or audit results should be added to the list of potential improvement opportunities.

<u>Recent High-Profile Incidents</u>

Consider placing on your list for study any external incidents that involve a technology, process, or unit operation the company uses. The findings and recommendations from the investigations of those incidents may be directly applicable to company standards, policies, and PSMS. In the special case of licensed technology, the technology licensor may do some of the evaluation on behalf of all licensees. However, don't assume the licensor has done the full evaluation. The company is ultimately responsible for its own operations.

<u>Due Diligence Evaluations and Post-Acquisition Integration</u>

Process safety evaluations conducted during and following acquisition of a site typically identify gaps that need to be closed relative to the company's PSMS, standards, and policies. If the operation of the business, technology, or unit differs from what the company practices, it is worth studying any relevant external incidents. Similarly, if the company's PSMS, standards, and policies don't provide a ready resolution for the identified gaps in the acquired business, external incidents may provide some guidance.

New Process Development and Design

New processes may introduce process safety hazards or management challenges that the company is unfamiliar with. Place on the list for study any external incidents relevant to the new process. This may not only inform design but also ultimately the commercial implementation of the process.

New Product or Technology Strategy

Whenever a product or technology strategy changes, the company may encounter new process safety hazards and management requirements. How you address them may be informed by external incidents. Here are some examples of changes that could introduce new hazards:

- changing from a single-product train to multi-product train.
- replicating an existing process in a country with different process safety requirements
- replacing one licensor's technology with another
- changing from production-to-inventory to production-on-demand.

When considering changes of this type, place any relevant external incidents on the list for further study.

Anticipated Downsizing, Growth, or Reorganization

If there is one thing we can count on in this industry, it is that organizational structures will change. Any time there is a change in organization structure, critical roles may go unfilled or be under-resourced. In this situation, it makes sense to learn from as many incidents related to organizational change as possible and use this learning to guide the organizational change process.

6.1.2 Choosing Learning Goals and Priorities

Leadership is responsible for setting priorities for learning from incidents. To do this, leaders and process safety specialists should consider the learning drivers discussed in the previous subsection relative to the three types of learning triggers summarized in Table 6.1.

Table 6.1 Possible Triggers for Seeking Learning

Trigger Type	Learning Trigger
Regular, ongoing learning	• employee development • required regulatory training
Events and conditions	• high-profile incidents • acquisition and post-acquisition integration • new process development and design • new product or technology strategy • organizational changes or new initiatives
Metrics and data	• tier 4 leading metrics • audit findings and recommendations • tier 3 near-miss metrics • incident and near-miss investigations • gut feel and warning sign analysis

In general, give the highest priority to learning opportunities that metrics and data indicate will have the largest positive impact on process safety performance. These generally come from Tier 4 and Tier 3 metrics. To help prioritize, think of the worst thing that could have happened if the Tier 3 or 4 event proceeded to the worst-case scenario.

As learning opportunities triggered by events and conditions arise, leaders should fit them into the priority list as appropriate. Finally, ongoing learning related to organizational issues should always be included to help the company react nimbly to change without compromising process safety.

Here's a hypothetical example. During a management review meeting, the leadership team of a refining company discusses the points presented in Table 6.2 below. From this, they develop corresponding short-term improvement goals.

Table 6.2 Example of Process Safety Goal Development

Discussion point	Possible Goal
In the past year, seven facilities had audit findings related to operating procedures.	Deploy a common operating procedure based on best practices and audit to ensure it is followed.
In the past two years, the Tier 4 metrics for asset integrity ITPM have slipped from 99.5% completed on schedule to 97%.	Evaluate gaps in the asset integrity program and recommend actions for improvement.

Table 6.2 Example of Process Safety Goal Development (Continued)

Discussion point	Possible Goal
A refinery owned by a competitor had a serious release and explosion at a unit using the same licensed technology the company uses at most of its refineries.	Evaluate the competitor's incident and determine whether any of the root causes or causal factors apply to the company's facilities. Implement corrective measures as needed.
At their next turnarounds, two facilities will convert their alkylation catalyst from hydrogen fluoride to sulfuric acid.	Corporate HIRA and MOC programs are deemed adequate to address this change. No learning goal needed.
Analysis of near-miss data shows that in the past year, there has been a spike in tank high-level alarms.	Determine the cause of high-level alarms and recommend actions for improvement.

In assigning personnel in alignment with these goals, the company process safety leader should work with corporate leadership to set the priority for examining findings and recommendations from incidents. One possible prioritization based on Table 6.2 could be:

1. *tank overflow.* Since overflow is a loss of containment, leaders fear that, without prompt action, such an incident could happen soon.
2. *asset integrity.* Since asset integrity failures often result in losses of containment, leadership assigns the next highest priority to this topic.
3. *operating procedures.* Standardizing procedures represents a key investment in operational discipline. With key strategic value, leadership places this third in priority.
4. *competitor's incident.* It will take some time before the findings from the public investigation will be released. So, this item is set aside for the present. But when findings are made public, leadership knows this item will move up in priority.
5. *technology change.* This is given the lowest priority. However, leadership makes a point of discussing relevant external incidents with the technology licensor and engineering company to ensure that any findings from those incidents have been addressed in the design, HIRA, and MOC.

Each reader and leadership team might make different prioritizations than just described. That is perfectly acceptable. The order in this example was colored by the authors' experience, and each reader's experience is almost certainly different. The bottom line is that any effort to learn from external incidents should be based on company goals and priorities. This will help deliver the greatest value for the effort.

6.2 Seek Learnings

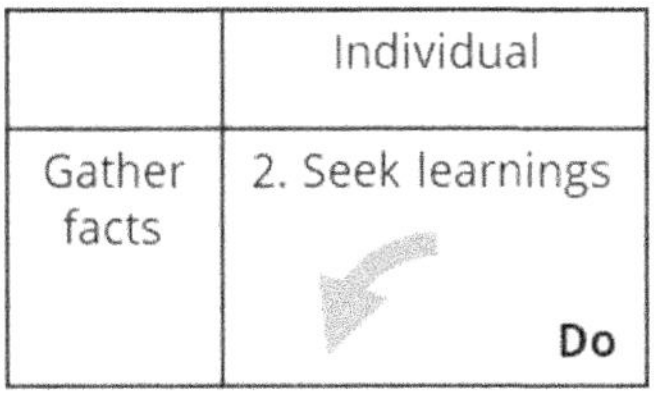

Having been guided by the company's process safety improvement goals, the evaluator seeks internal and external incident reports that have the potential to provide important learning.

	Individual
Gather facts	2. Seek learnings **Do**

How do you identify the most relevant incidents external to your company? Start with the Appendix of this book. You can search for publicly reported incidents based on the management system elements and culture core principles that appear to have failed in each incident. You can also search on selected causal factors. Additionally, the electronic version of the index enables you to search based on type of equipment and industry, and you can search on multiple parameters. These features allow quick identification of incidents with findings relevant to the improvement goals.

Other free public sources of incident reports exist, including those mentioned in Chapter 2. Conference and journal papers can be obtained, sometimes for a small fee. A simple Internet search may also reveal additional useful case histories. Some countries have a policy of retaining government investigation reports outside of the public domain but provide ways to request copies.

If the report is not in your primary language, various free translation services available online are generally are good enough to allow you to decide whether it would be valuable to obtain a more formal translation. Finally, though many incident reports are in text format, several public investigation organizations produce reports as videos and screen presentations (such as PowerPoint®).

Once you have identified the most relevant incident reports to study, find a place where you can maintain a steady focus, get comfortable, and read.

6.3 Understand

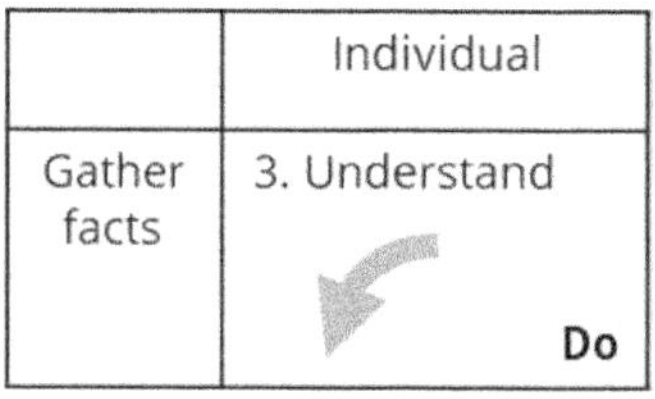

	Individual
Gather facts	3. Understand Do

While reading incident reports, you should flag the text that pertains directly to corporate learning priorities. Also flag other items of interest that might merit later follow-up. These might include observations that trigger obvious to-do items, or that address lower or potential future priorities.

Try to imagine yourself in the plant before the incident. What was happening in the weeks, days, and hours before the incident? How were decisions being made? What would you have heard or seen if you were there? How might that information have influenced the way you think about the incident? Ultimately, you need to become intimately familiar with the details of the incident, essentially making a mental model of how the incident occurred and how it might have been prevented.

Not all incident reports contain the same amount of detail. If you find that an incident report might have left out a critical piece of information, note what's missing for consideration in the next step.

Take note of commonalities between the incidents and think of how the conditions in those plants compare to your own. It would not be unusual to look at an external incident report and find your mind beginning to dismiss what happened as something your company would never do. If you find your thoughts turning in this direction, remember that personnel at the company that had the incident were probably having similar thoughts right before the incident happened.

6.4 Drilldown

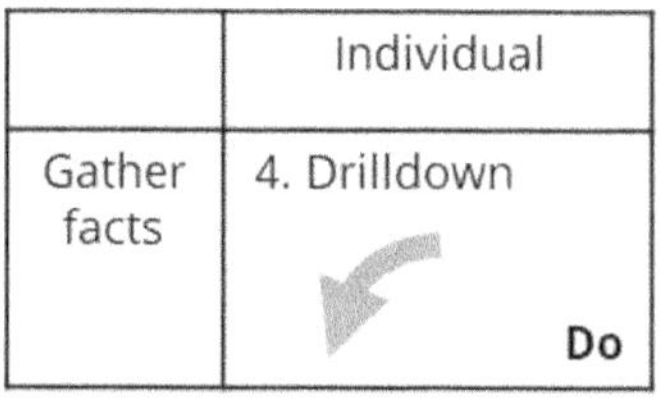

	Individual
Gather facts	4. Drilldown Do

In this step, the evaluator intentionally goes beyond the boundaries of the incident report to discover any deeper learnings that might not be explicitly stated in the investigation reports. This is important because the investigator may have had different learning objectives than those driving the evaluator's study.

The goal is to develop a list of deeper learnings from the relevant external incident that are tailored to your plant or company. Part of getting to deeper learnings involves asking, "Did the investigator…":

- *Miss something important*? For example, in their Moerdijk investigation, the Dutch Safety Board (DSB) noted that:

  …an automatic protection system was triggered that was designed to prevent liquid from entering the…flare. As a result, the gases in the system were no longer able to be discharged (DSB 2014).

  DSB then focused their findings on the initiating event (that is, an unexpected reaction) and didn't discuss the fact that the interlock designed to prevent liquid carryover blocked the pressure relief system, a critical barrier that may have prevented the consequences if it had been able to function.
- *Fully understand what they observed*? For example, in their investigation of the Texas City refinery explosion, CSB observed:

  The tower level indicator showed that the tower level was declining when it was actually overfilling… (CSB 2007).

  and described that condition as a malfunction. However, given the physical principles of differential pressure tower level indicators, any differential pressure indicator would have shown the same behavior if the level rose beyond the upper pressure tap. As the level increases, pressure and boiling point temperature increase, causing density to drop. That makes the level appear to decrease, which can easily fool operations personnel.
- *Arrive at causal factors, root causes, findings, and recommendations that differ from what you would have concluded*? Considering the above two possibilities, you shouldn't be surprised that you will see something different than the investigators did. And if you do, your thought process is probably colored by your experience in your company's plants. While the original investigators' findings are still valid, the alternative or additional root causes, causal factors, findings, and recommendations you might find are also important in the next step of the process.

Also, consider the missing information you identified in the previous step. In some cases, you can contact the investigating organization to request additional details. You may be able question a member of the investigation team or obtain investigation notes. You may even be referred to a contact person at the company so you can ask them directly.

Finally, whether it's better, worse, or only different, your plant almost certainly has different design features, preventive and mitigative barriers, and PSMS, standards, and policies than the company that had the incident. So, think about how the incident scenario might have played out if it happened in your facility. What warning signs might have been present pre-event? Would your systems have prevented the incident? If so, how do you know they would have functioned reliably?

6.5 Internalize

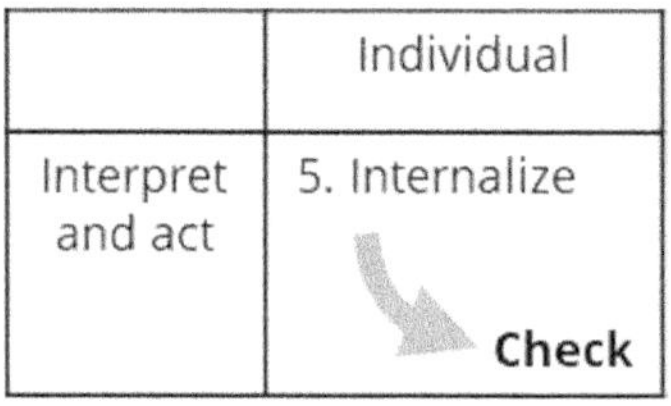

	Individual
Interpret and act	5. Internalize **Check**

The next step is for the effort to expand beyond the individual evaluator to include a small team of diverse individuals who provide relevant expertise. In this step, the group evaluates the list of deeper learnings and internalizes them to the company by developing formal recommendations. This step is important because the recommendations developed for the external incident may not work for your company given its expertise, resources, or technologies. Alternately, your company may have access to better solutions. And finally, a solution that comes from a company's experience and culture is most likely to be accepted and become institutionalized.

Depending on the subject matter, the team members will include those who conducted the initial review as well as others with expertise in relevant areas, such as:

- process technology
- engineering
- corporate and public standards
- HSE policy
- process economics and finance
- manufacturing and operations
- procurement
- human resources
- transportation and logistics.

Ideally, the team will also include one or more individuals who will ultimately be responsible for implementing any changes that result from the team's work.

The team develops recommendations addressing the corporate learning objectives, just as a team investigating an internal incident or near-miss would. The team may also consider developing additional recommendations that address findings of interest from external incidents. CCPS covers the recommendations process in detail (CCPS 2019b). In summary, the recommendations must:

- be SMART, that is, specific, measurable, attainable, realistic, and time bound
- address the corporate improvement goals developed in Step 1
- address specific root causes and/or causal factors
- result in improvements to one or more PSMS elements, standards, policies, or business practices that apply broadly
- map the action to the learning and demonstrate how risk will be reduced. Ways to reduce risk include:
  - correcting an ineffective barrier
  - reducing potential consequences
  - reducing the probability of occurrence
  - a combination of the above.

The team may offer more than one recommendation addressing any given corporate improvement objective. The recommendations may be either alternative, additive, or both. Alternative recommendations offer multiple options for addressing a given improvement objective; after evaluation, the best can be selected. Additive options work together, with each option moving the company closer to meeting an improvement objective.

6.6 Prepare

	Individual
Interpret and Act	6. Prepare **Check**

In this step, recommendations become action plans. These may involve one or more of the following:

- changes to a policy
- changes to a standard
- capital expenditures
- new risk-reduction measures

- new or revised business processes
- changes in ITPM or operating procedures
- new hires, and/or realignment of responsibilities
- training and ongoing communication.

The individuals responsible for preparing action plans may or may not be the same ones involved in making the recommendations, depending on the skill sets required. However, the individual responsible for the evaluations in steps 2–4 will likely remain involved.

For recommendations to have the greatest weight, consider preparing and presenting action plans in the format of completed staff work (US Army 1942). That is, do the complete professional and economic analysis to be able to present all the information that the approving leader will need to authorize the action. This includes:

- the reason for the change (e.g. REAL Model Step 1)
- the problem to be solved (e.g. Steps 2– 4)
- the actual work recommended (e.g. Step 5) and why it is best
- other alternatives considered (also Step 5) and why they were rejected
- additional benefits expected
- how the plan will be implemented
- costs, resources required, and timing.

A business plan for any other initiative would contain this information; process safety demands the same level of rigor and detail.

The action plan should be sensitive to other business priorities when feasible. For example, if the plan involves capital improvement that would require plant shutdown, and a turnaround is expected in nine months, it may make sense to implement the plan as part of the turnaround, if safety would not be compromised by waiting.

Plans addressing changes to a standard or policy should address the cost of bringing equipment and procedures up to the new standard or policy. Plans addressing changes to the PSMS should address any changes to supporting business systems and processes.

Finally, the plan should include measures to ensure that change is sustainable. These may include an ongoing communication plan, changes to management review and conduct of operations expectations, and updates to audit protocol.

In most cases, it will help to discuss the draft plan with key stakeholders before the formal presentation. It's important to know what, if any, questions or objections they may have and address these in the final proposal and presentation. Stakeholders may also be able to offer insights into how their peers will receive the proposal and how to win them over.

The action plan will usually be presented to the same group that set the improvement priorities laid out in REAL Model Step 1. Plant or business-specific plans are generally presented to the leadership team for the plant or business, while corporate level plans are generally presented to the senior leadership team or Board of Directors.

6.7 Implement

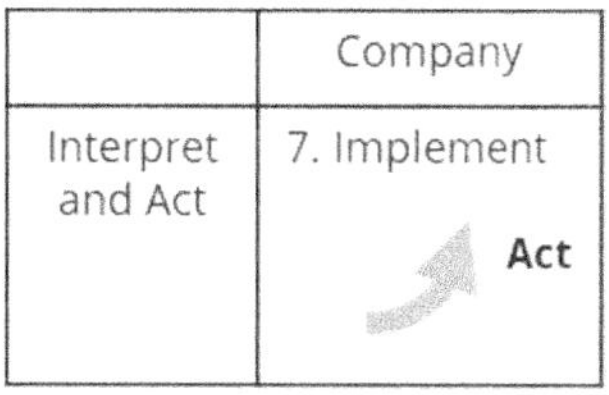

	Company
Interpret and Act	7. Implement **Act**

Once the plan is approved, it should be implemented on the agreed-on schedule. The makeup of the implementation team will depend significantly on the type of change being implemented.

Regardless of the type of change, senior leadership must show their support and involvement. For corporate-wide changes, support should come from both corporate leaders and plant leaders. Support goes beyond writing an email or giving a speech. Depending on the change, leaders may choose to take steps such as:

- visibly commit to the change, including commitment of financial and human resources
- demonstrate the new behaviors expected
- follow up on progress and/or monitor compliance in management review meetings or conduct of operations activities
- visit the site where the change has been implemented and engage with workers
- celebrate success with an employee gathering.

Implementation almost always requires additional PSMS activities such as MOC, training, updating procedures, operational readiness/pre-startup safety review (PSSR), and updating audit protocols. In every case, it will be essential to update the process safety knowledge base to reflect all the changes made, along with the reasons for the change. For changes arising from using the REAL Model, the case studies and their analyses should also be added to the knowledge base. Ideally, these case studies will be described in variety of ways that consider different learning styles (see Section 5.2.1; more details are provided in Chapter 7).

6.8 Embed and Refresh

Company and site leadership must now manage the changes as implemented. Assuming the PSMS has been updated, much of the work of maintaining continuity of the change will happen via routine conduct of operations and management review activities.

	Company
Interpret and Act	8. Embed and Refresh **Act**

However, experience has shown that without regular reminders, including ongoing verification of performance, the organization will gradually forget the reason for the change. Normalization of deviance will then set in, the sense of vulnerability will diminish, and ultimately the knowledge will be forgotten. Strategies for providing plant and corporate personnel with regular reminders of key lessons learned will be presented in the following chapter.

6.9 References

6.1 API (2016). Process Safety Performance Indicators for the Refining and Petrochemical Industries. ANSI/API RP 754 2ND ED (E1).

6.2 CCPS (2007). *Guidelines for Risk Based Process Safety*. Hoboken, NJ: AIChE/Wiley.

6.3 CCPS (2011a). *Guidelines for Auditing Process Safety Management Systems*, 2nd Edition. Hoboken, NJ: AIChE/Wiley.

6.4 CCPS (2011b). *Recognizing Catastrophic Incident Warning Signs in the Process Industries*. Hoboken, NJ: AIChE/Wiley.

6.5 CCPS (2016). *Guidelines for Integrating Management Systems and Metrics to Improve Process Safety Performance*. Hoboken, NJ: AIChE/Wiley.

6.6 CCPS (2018a). *Essential Practices for Building, Strengthening, and Sustaining Process Safety Culture*. Hoboken, NJ: AIChE/Wiley.

6.7 CCPS (2018b). *Process Safety Metrics: Guide for Selecting Leading and Lagging Indicators*, Version 3.2. New York: AIChE.

6.8 CCPS (2019a). *Process Safety from the Boardroom to the Frontline*. Hoboken, NJ: AIChE/Wiley.

6.9 CCPS (2019b). *Guidelines for Investigating Process Safety Incidents*, 3rd Edition. Hoboken, NJ: AIChE/Wiley.

6.10 CSB (2007). BP America Refinery Explosion. Report No. 2005-04-I-TX.

6.11 DSB (2014). Explosions MSPO2 Shell Moerdijk. Dutch Safety Board report.

6.12 Rae, S. (2013). A Survivor's Experience on Piper Alpha [Video]. www.youtube.com/watch?v=1wNG3LfEg6o. [Accessed June 2020)

6.13 US Army (1942). The Doctrine of Completed Staff Work. govleaders.org/completed-staff-work.htm. (Accessed January 2020).

7
KEEPING LEARNING FRESH

"Learning is like rowing upstream: not to advance is to drop back."—Chinese Proverb

Researcher Barry Throness investigated how long employees remember the lessons learned from major incidents. He found that after an incident, employees remained attuned to the lessons learned and considered them in their work for about three years (Throness 2013). After that, as the memory of the incident began to fade, employee behaviors reverted to what they had been before the incident. With regular and frequent reminders of the incident, however, compliance could be maintained.

A study of learning unrelated to process safety, by researchers at the Wharton Business School, showed a similar result (Michel-Kerjan 2012). In 2006, the year after Hurricane Katrina's massive flooding in New Orleans and the US Gulf Coast, the National Flood Insurance Program experienced a 14.3% increase in new US flood policies, three to four times the normal growth rate. After 2006, however, the policy cancellation rate increased as homeowners became complacent. The study also showed that, in general, a certain fraction of policy holders drop their insurance each year; 67–77% hold their policies longer than one year, but only 29–44% hold their policies longer than five years.

Studies of memory are nothing new. In the late nineteenth century, Hermann Ebbinghaus conducted experiments involved reading random syllables and then testing how long it took for these random syllables to fade from memory (Murre 2015). After 25 years of testing his own memory, he published what he termed the forgetting curve, empirically described by the equation:

$$R = e^{-t/s}$$

where R is the fraction of memory retained, t is time, and s represents how stable the memory has become. Many subsequent researchers have shown that memory can be made more stable with (1) frequent repetition of the lesson soon after the first learning, followed by (2) less frequent, but regular repetition over time. Both are critical.

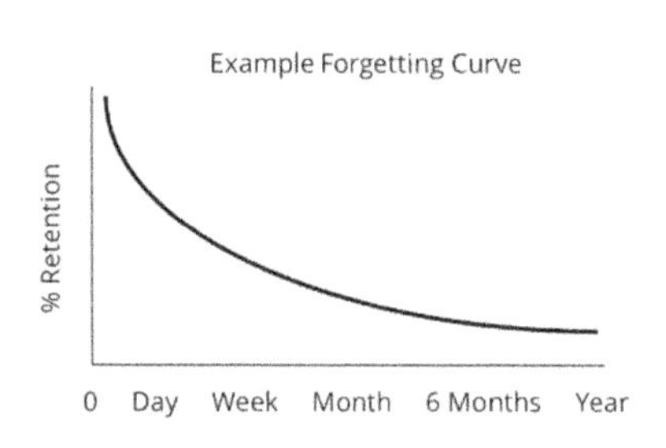

In the Throness study, employees clearly were receiving the first part of the process of memory retention. After any incident in a facility, the lessons learned will initially be repeated frequently, as the plant goes through the process of rebuilding and restarting operation. Similarly, in the Michel-Kirjan study, the experience of living through Hurricane Katrina and then rebuilding provided the initially strong reminders. But then, the reminders stopped as things returned to "normal," and the lessons were forgotten.

Then, what about employees at other company plants, or people who live in hurricane-prone areas away from where Katrina hit? They may not have received the initial strong reminders. For many them, the lessons-learned may be fleeting.

Therefore, in implementing the REAL Model, special attention must be paid in the Embed and Refresh step to communicate the lessons-learned frequently at first, and then follow up with regular reminders.

In Section 5.2.1, we discussed the theories of multiple intelligences and learning styles. We concluded that, because everyone can expect to see a broad range of learning styles among members of their workforce, we needed create tools for learning from incidents that are suited to many of these styles.

In the following sections, we will discuss ways that companies and other organizations can create institutional knowledge using the REAL Model—and keep it fresh by communicating in ways that consider Gardner's theory of multiple intelligences. We describe ways to use different communication styles, providing both hypothetical and real-world examples from committee members' experiences at a range of companies.

We will also show how communicating in multiple ways can reinforce one important process safety message that is often forgotten in the heat of the moment: Resist the very human urge to rescue a colleague who has collapsed (CSB 2008) and call for trained responders instead. We know intellectually that, if we see someone collapse in a hazardous work environment, the cause is much more likely to be a toxic or asphyxiant exposure than some medical issue such as a heart attack. Despite this knowledge, people often respond instinctively by trying to help and become victims themselves. Everyone must be reminded frequently that in a person-down emergency, the right step is to call for appropriately trained personnel who can assess the hazards and respond safely using the proper personal protective equipment (PPE).

Regardless of which learning style your employees favor and which teaching strategies you use, it is important to test following the learning experience. This helps assure that the message you meant to convey was received by the learner.

7.1 Musical Intelligence

Gardner discusses musical intelligence first, perhaps because music is basic to the human experience and at the core of the brain (Gardner 1995). And even after Alzheimer's disease has robbed people of their ability to remember people and events, many can still sing songs they learned years earlier (Doward 2014).

For this reason, music is used widely in popular media—film, theater, commercials, etc.—and in training videos. The US Chemical Safety and Hazard Investigation Board (CSB) uses music in many of its videos to help reinforce the seriousness of the messages they wish to convey. Music similarly plays an important role in the BBC's special program about a process safety incident, *Piper Alpha, Spiral to Disaster* (BBC 1998). Whenever viewers hear music, they feel a sense of vulnerability to the next development in the sequence of events. Additionally, background sounds in the video helps viewers feel the rhythm of the rig, as if they were right there in the moments leading up to the tragedy. This further builds the sense of vulnerability.

Another use of music is to deliver a parallel message. In the background of their video "Walk the Line' (CCPS 2017), employees of the Celanese Congreja, VE, Mexico plant play the song "Happy Together" (Gordon 1967) as they show an animation of a worker being instructed how to make sure all valves are in the right position before starting a transfer. Even though an instrumental version of the song is used, the song is sufficiently popular worldwide that viewers understand the message that "me and you, you and me," must work together for process safety.

Some industries have a long tradition of songs about tragedies, most notably the mining and maritime industries. So why not process safety, too? To date, the authors know of only one song composed in a plant about a process safety incident, "The Wreck of the [Product Name] Spray Dryer" (Anonymous). The song, describing an actual dust explosion, was set to the tune of "The Wreck of the Edmund Fitzgerald," a popular song about a 1975 shipping disaster (Lightfoot 1976).

Writing a song or a rap about a process safety lesson forces you to distill the consequences, root causes, and necessary actions very concisely. Industrial tragedy songs typically follow the format of public incident investigation reports, starting with the first inkling there might be a problem, then describing the timeline and the impacts on the victims and survivors. To demonstrate this, the authors developed a song about the Bhopal tragedy. Perhaps the shortest piece ever to be written about Bhopal, it does contain an important lesson to be learned.

Bhopal's Lesson About Barrier Maintenance

Set to the tune of The House of the Rising Sun (Price 1964)

Back in 1984, a city called Bhopal,
A tragedy unfolded, one midnight in the fall.

The flare was down for maintenance, the chiller's freon went,
The run-down tank full of MIC, the scrubber caustic spent.

All barriers out of service, when tank 610 ran away,
40 tons of MIC spread into the night of gray.

Two thousand people died that night, thousands more for years,
Whole families perished, none left to count the tears.

Friends, tell your colleagues, how to avoid this strife,
Test and maintain your barriers, and they will save your life.

Back in 1984, a city called Bhopal,
A tragedy unfolded, one midnight in the fall.

Beyond learning from composing a song yourself, another use of music is to serve as a simple reminder of the message. See the REAL Model scenario in Chapter 11 for a safety jingle composed as one way to continuously reinforce the message the team wished to convey.

To communicate musically about responding to person-down incidents, you could create a short video that uses music to reinforce the message. Picture this: As a worker approaches his downed colleague, you hear the soundtrack from a popular movie that tells you something bad is about to happen—perhaps the shark's theme from *Jaws* or Darth Vader's theme from *Star Wars*.

Whether you use music to help reinforce the message, enhance a sense of vulnerability, add a parallel message, or describe the incident itself, it can be part of the lessons learned reminders you deliver to your company.

7.2 Visual-Spatial Intelligence

Many public incident reports are summarized in video form. Videos are particularly useful for helping viewers visualize how the incident occurred: the process flow, the spatial arrangement of piping and vessels, and the evolution of the incident through the equipment. Some companies also create "Look Back in Time" videos, in which leaders discuss past incidents that influence how the company currently operates.

Animations generally work better than actual videos of the incident or the facility because they focus viewers' attention on the key points, removing details that might otherwise distract from the message. The CSB is well known for using animations to help show how incidents evolve. Another good example of animation for this purpose was produced by the Brazilian oil and gas company Petrobras, depicting the events leading to the explosion and sinking of their oil platform P-36 (Petrobras 2015). This video, presented in Portuguese with English subtitles, is effective even if you don't speak these languages.

Live footage is better than animation for demonstrating the consequences of an incident, however. It helps enhance the message about "Why we need to do this." Two of the most effective real-time video segments of actual incidents are from the CSB St. Louis investigation (CSB 2006) that shows exploding gas cylinders rocketing through the air and the Reynosa surveillance video (Reynosa 2012) showing the evolution of a gas line rupture, followed by a flash fire and jet fire. A caution here: If you use live footage as a teaching tool, avoid up-close, graphic depiction of injuries. At the sight, many observers will leave the room—either physically or emotionally—and that undermines the message.

Another visual-spatial technique is a drawn diagram. T.J. Larkin, a well-known safety communicator, advocates hand-drawn diagrams over photographs for highlighting the issue for the same reason that animation is preferred over actual video of an incident–their simplicity makes it easier to focus on the important details (Larkin 2012). Repsol provided an example of such a diagram addressing a utility service contamination event, an excerpt of which is reproduced as Figure 7.1.

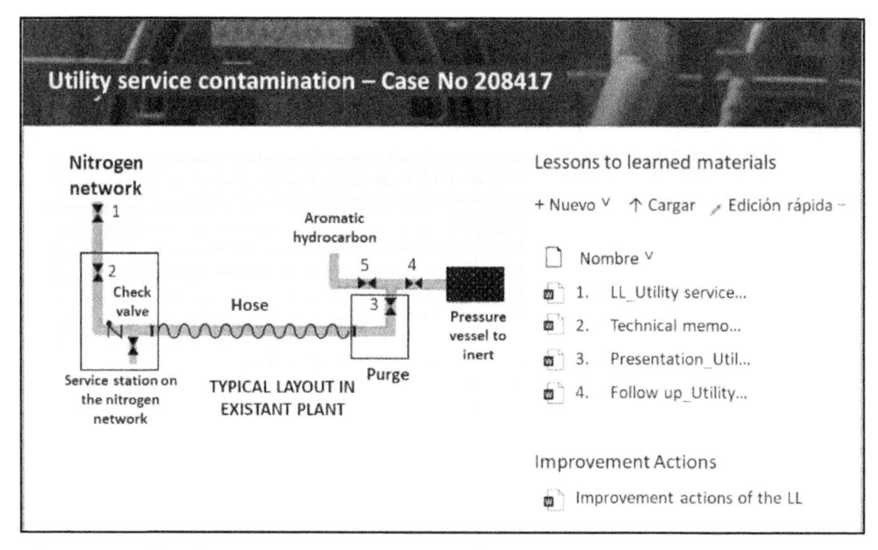

Figure 7.1 Simple Drawn Diagram Example (Source: Repsol, reproduced with permission)

Finally, it can be remarkably effective to leave evidence of past incidents in place to serve as a daily reminder of why we must do what we do. One company left shrapnel from a significant explosion lodged in a wall, with a sign commemorating the event. Another company has a framed photograph showing the plant on fire hanging by the employee entrance, with the caption, "Never again. We all know what we must do." The learning model scenarios in Chapters 9, 10, and 11 use posters with graphics and/or videos to stress the importance of proper safety practices.

To communicate visually about responding to person-down incidents you could create a video, as mentioned in Section 7.1. Alternatively, you could draw a simple poster, such as shown in Figure 7.2 (facing page).

Figure 7.2 Simple Poster About Response to a Person-Down Incident

7.3 Verbal–Linguistic Intelligence

Written incident reports and bulletins are the most common ways that findings from process safety incidents are captured and communicated. Writing the report may be as instructive to the members of the investigation team as it is to the people who later read that report.

Larkin suggests several techniques to improve verbal-linguistic learning (Larkin 2018):

- *Write in simple, direct sentences*. This can be measured via an option in MS Word spell-check. Target a Flesch-Kincaid reading level of 8 if possible.
- *Keep line length short*. The 4.5-inch line length in this book is slightly longer than the 3– 4 inches Larkin recommends.
- Keep paragraphs short. Short paragraphs hold readers' attention better than long ones.
- *Use a sans-serif font*. The font in this book is one example. Serif fonts can reduce the efficiency of learning from written messages.

Example font types

Sans-serif font

Serif font

Many companies organize their process safety learning bulletins on intranet sites. Figure 7.3 shows an example.

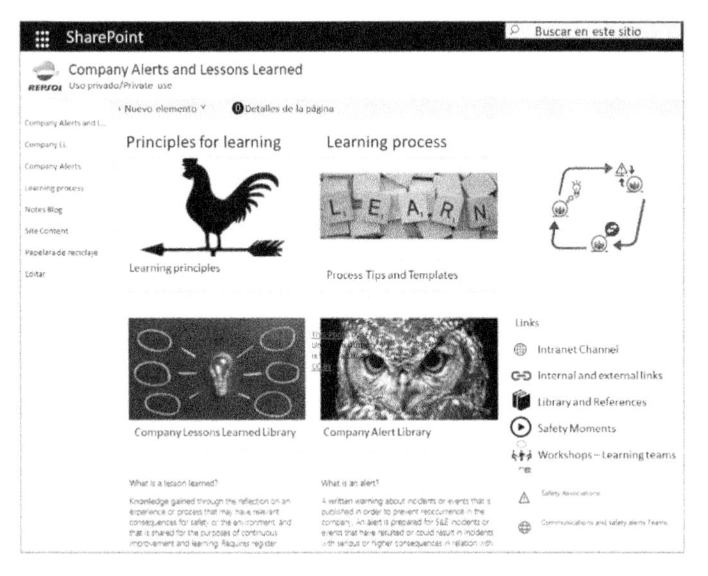

Figure 7.3 Lessons learned home page (Source: Repsol. Reproduced with permission.)

Conveying messages verbally is another way to reach people with strong verbal-linguistic intelligence. Verbal communications may have increased impact when a supervisor or other trusted influencer is the one sharing information about an incident and its findings and recommendations. The challenge with verbal communication is the possibility that the message can evolve with time. For this reason, verbal communication of key process safety messages should be standardized and monitored for consistency.

Because not every employee is oriented to verbal-linguistic learning, many companies seek to distill lessons learned into concise written communications that are easier to process. For example, one company effectively communicates essential learnings about how to prevent distraction and encourage situational awareness using simple pocket cards and matching posters (Anonymous). Figure 7.4 shows an example of a pocket card. The pocket card goes into the badge holder, so the employee sees it whenever they use

[Company name and logo]

- Eyes on the task
- Mind on the task
- Out of the line of fire
- Do it safely or don't do it

Figure 7.4 Simple pocket card

their badges to enter or exit. The posters are distributed widely throughout their facilities.

The act of writing or telling about incidents and the key lessons learned from them is itself a way of reinforcing the message. Some companies organize a poster or video contest inviting employees or teams to develop communications based on lessons learned. The individuals who participate learn (or are reminded) in the course of their hands-on work, and the products they develop are also useful for those attuned to visual-spatial or verbal-linguistic learning. Nearly all the learning model scenarios presented later in this book (Chapters 9–14) mention this mode of communication. The company in Chapter 9 uses a memorable and concise sign to call attention to the company's safety practices.

There are a variety of ways to communicate verbally/linguistically about the proper response to person-down incidents. A natural time to remind every employee is on the anniversary of a person-down incident, whether it happened at your company, at another company in the community nearby, or in some other location that would be meaningful to employees. The reminder could take the form of an email, a public address announcement, or a direct verbal message, for example at a sitewide or unit-wide meeting. Describe how the incident occurred. Ask employees to take a moment to remember the victims. Then remind them of the key learning: You will likely be harmed if you try to rescue a coworker. So, if you see a person down, call for help.

7.4 Logical–Mathematical Intelligence

The logical–mathematical style is frequently used in incident investigation. Causal trees, fault trees, and event trees help investigators reason through the hierarchy of causes and arrive at the root causes and causal factors. Such trees typically are then reproduced in investigation reports.

People with the logical-mathematical style of intelligence type will benefit from delving into the details of how the incident progressed, from root causes and causal factors through intermediate failures to the ultimate incident. If the details and intermediate steps are omitted, the message may seem to them to be of less value or seriousness. It is important to provide a way they can understand all the details. The learning model scenario in Chapter 10 offers an example of effective communication using logic. It features a poster that lays out the sound logic of standing on floors, ladders, scaffolds, or person-lifts rather than on piping, hoses, conduits, or equipment.

You could also create your own learning exercise keyed to this style of communication using readily available short videos, for example the CSB video about the Illiopolis, IL, USA incident. Show the video (CSB 2007) and then ask participants to break out into groups to answer these questions:

- Why did the operator use the emergency air hose without thinking?
- Why did the supervisor run upstairs to stop the release rather than evacuating?
- For extra credit: Why was a hose clearly labeled "Emergency Air" provided under each reactor?
- Based on your answers to these questions, how would you ensure your team acted properly?

During the report-back time, steer the conversation toward the underlying issues, including normalization of deviance, conduct of operations, human factors, and where in the PHA this scenario might have been captured.

Here's another way to communicate in a logical-mathematical style about the proper response to person-down incidents. Create a bulletin, poster, video, or presentation showing that the incidence of health conditions like heart disease and diabetes is lower among workers in the process industries than in other industries. If possible, compare your company's own statistics for these conditions to the rate in the general population. Then ask, "In a facility like ours that handles toxic materials and asphyxiant gases, why would the person have collapsed on the floor? What should you do?"

7.5 Kinesthetic Intelligence

People with kinesthetic intelligence learn best when they can have hands-on learning experiences. To support this kind of learning, aspects of some incidents can be simulated under safe conditions: for example, controlled demonstrations of chemical reactivity, vapor cloud, and dust explosions. Simulations are especially effective when observers can feel some heat from the fire or feel and hear the pressure wave from the explosion. Computer simulations can also be effective learning tools, although it is harder to simulate the heat or pressure wave. Role-playing is another a beneficial way to learn, especially if a moderator can supply stimuli and sound effects that help the participants imagine what an explosion or fire feels like.

Another aspect of kinesthetic learning is learning while moving. After a major incident occurred in its facility, one company set up a walking path with signs along the way depicting how the incident unfolded. As workers walk the path, they learn how the incident occurred and how to prevent it. The learning model scenario in Chapter 14 uses this method of learning with a hands-on contest where the challenge is how to handle a simulated ammonium nitrate incident.

An inexpensive and fun way to help colleagues appreciate fire and explosion hazards is to take them to dinner at a Japanese steakhouse. After the chef performs the ritual of igniting a few milliliters of 190 proof alcohol (a practice that has been determined by experience to produce a burst of flame that makes patrons uncomfortable but doesn't hurt them), ask him to share with the group what quantity he used. Then have the participants guess the volume of flammable materials in a typical industrial spill and how far they would need to be from the deflagration in order to be safe. After dinner, walk with the team away from the restaurant, stopping at the distance where the radiation from the hypothetical fire would have dissipated to the level they felt during the indoor demonstration. Participants will likely be surprised how far away they had to walk.

Here is another fun way to promote kinesthetic learning of the proper response to person-down incidents. While you are training a group or leading a meeting, collapse to the ground at a random time, pretending to be overcome. If you are indoors, stand near a source of outside air such as a vent. Before collapsing, draw attention to the vent by pointing at it and asking the group, "Do you smell that?"

When someone rushes to help you (for the first person, you could enlist a helper in advance), hand them a card that looks like the text box to the right. Depending on how well employees understand person-down procedures, you may need to have several cards ready to hand out. After an appropriate number of mock fatalities, revive yourself and discuss with the group what everyone should have done.

You're dead. Fall down. Hand this card to anyone who tries to help you.

7.6 Interpersonal Intelligence

People with interpersonal intelligence learn best where they can discuss the lessons learned with a group. This fits well with the format of safety meetings and toolbox talks. It can also fit well as part of safety moments at the start of regular business meetings, providing that the moderator engages the participants and doesn't simply lecture.

Interpersonal learning can happen in tandem with most of the communication styles discussed in this section. Groups can read reports or watch videos and then discuss them, they can collaborate on written materials, videos, or even music, and they can discuss lessons learned and how they apply to the facility during a group walk-through. Several of the learning model scenarios provided later in this book (Chapters 9–14) use this form of learning. In Chapter 10, one possible way of engaging employees is to ask them to create a skit to allow for role-playing, then initiate a discussion. In Chapter 11, the scenario involves toolbox talks and safety moments at the start of each shift.

Some companies form formal operational learning teams to address process safety issues. While the issues could probably be resolved easily by process safety professionals, involving teams turns the exercise into more of a learning process, and the team members take greater ownership of the result. To address person-down incidents by this style, ask teams across the facility to collaborate on how best to train for these incidents—and how to spread the word about best practices.

7.7 Intrapersonal Intelligence

People with intrapersonal intelligence benefit from time to think privately about the learning being conveyed and prefer reading or doing online training. If you've attended any learning event where there is an interpersonal discussion session, you will see individuals whose strength is intrapersonal intelligence sitting off to the side reading or thinking about the topic.

It's important to recognize this learning style and give people the space to learn in the way most comfortable to them. All the companies in the learning model scenarios (Chapters 9–14) make safety information available to their employees. Chapter 9 mentions using eLearning modules to refresh and reinforce safety practices.

As part of its zero-incident mindset (ZIM) initiative, the major global chemical company BASF provides one or two ZIM-grams to all employees each month (Miller 2020). Each ZIM-gram describes a near-miss or incident and highlights the lessons to be shared. While ZIM-grams are written for intrapersonal learning, they often have visual-spatial design elements and are frequently used in interpersonal settings such as safety moments at the beginning of leadership meetings.

Another form of intrapersonal learning uses games. To address learning related to person-down incidents in this way, the person-down exercise at the end of Section 7.5 could be turned into an online game.

7.8 Naturalistic Intelligence

Employees with naturalistic intelligence may view the process, PSMS, standards, and policies as comparable to living, breathing organisms. By extension they would see process safety as the care, feeding, and well-being of those organisms and view lessons learned in the context of process health. People whose strength is naturalistic intelligence receive messages better in a natural setting or when expressed using animals or plants. Figure 7.5 shows how the person-down message might be communicated effectively to these learners.

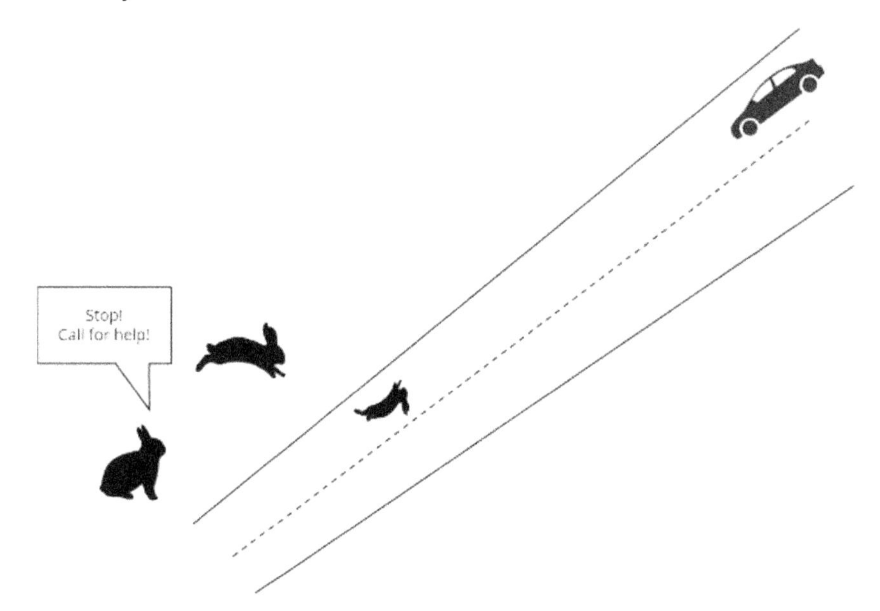

Figure 7.5 Example Naturalistic Communication About Person-Down

The poster created for the case presented in Chapter 9 uses a monkey to remind employees to not climb on equipment, piping, and hoses.

Naturalistic learners may not be as interested in numerical details as logical-mathematical learners but will still be interested in how the components of the system work together and how the lessons learned will help keep the overall system healthy.

7.9 Summary

Nearly everyone learns in multiple ways, but the mix of preferred learning styles varies from person to person. As you can readily infer from the examples in this chapter, many of the communication techniques described here can be combined. The important message is to remember to design communication approaches to support as wide a range of learning styles as you can. A lecture accompanied by text-rich slides will work for only a fraction of the workplace. So, spice up your communications with music or rhythm, role-playing, brainstorming, team activities, and the other techniques described in this chapter. And remember to provide individuals who need it the time for personal contemplation of the message.

7.10 References

7.1 BBC (1998). *Piper Alpha, Spiral to Disaster*. BBC Disaster Series.

7.2 CCPS (2017). Walk the Line [Video]. www.aiche.org/academy/videos/walk-line (accessed December 2019).

7.3 CSB (2006). "Praxair Flammable Gas Cylinder Fire." CSB Safety Video. www.csb.gov/praxair-flammable-gas-cylinder-fire/ (accessed April 2020).

7.4 CSB (2007). "Formosa Plastics Vinyl Chloride Explosion." CSB Safety Video. www.csb.gov/formosa-plastics-vinyl-chloride-explosion. (accessed April 2020).

7.5 CSB (2008). "Hazards of Nitrogen Asphyxiation." CSB Safety Video. processsafety.events/events/csb-hazards-of-nitrogen/ (April 2020).

7.6 Doward, J. (2014). Networked and super fast: Welcome to Bristol, the UK's smartest city. *The Guardian* (15 November).

7.7 Gardner, H. (1995). Reflections on multiple intelligences: Myths and messages. *Phi Delta Kappan* 77: 200–209.

7.8 Gordon, A. and Bonner, G. (1967) Happy Together [Song]. From the album "Happy Together" by The Turtles.

7.9 Larkin, T.J. and Larkin, S. (2012). Communicating Safety: The Silver Bullet. Larkin Communication Consulting.

7.10 Larkin, T.J., (2018). Communicating Safety. Speech to the AIChE Safety and Health Division (24 April 2018).

7.11 Lightfoot, G. (1976). The Wreck of the Edmund Fitzgerald [Song]. From the album "Summertime Dream" by Gordon Lightfoot.

7.12 Michel-Kerjan, E., Lemoyne de Forges, S. and Kunreuther, H. (2012). Policy tenure under the U.S. National Flood Insurance Program (NFIP). *Risk Analysis* 32 (4): 644–658.

7.13 Murre, J.M.J. and Dros, J. (2015). Replication and Analysis of Ebbinghaus' Forgetting Curve. *PLOS One,* 6 July 2015.

7.14 Petrobras (2015). Acidente da P-36 - Explosão e Naufrágio [Video]. www.youtube.com/watch?v=Oz10Rsw_bJc&t=5s (accessed May 2020).

7.15 Price, A. (Arrangement) (1964). House of the Rising Sun [Song]. From the album "The Animals" by The Animals.

7.16 Reynosa (2012). Gas Plant Explosion Mexico [video]. www.youtube.com/watch?v=6jhCKp2LHro (accessed April 2020).

7.17 Throness, B. (2013). Keeping the memory alive, preventing memory loss that contributes to process safety events. *Proceedings of the Global Congress on Process Safety*, San Antonio, TX (28 April–2 May 2013). New York: AIChE.

8
Landmark Incidents that Everyone Should Learn From

"We forget all too soon the things we thought we could never forget."
—From Slouching Towards Bethlehem *by Joan Didion*

This chapter discusses several landmark incidents that are critically important to our understanding of process safety. Everyone who works in a company that handles process safety hazards—leaders included—should have them engrained in memory and keep them in mind while doing work and making decisions. These landmark incidents have resulted in thousands of combined fatalities and injuries, severe damage to the environment, billions of dollars of loss, unwanted publicity, and a loss of the public's trust. Consider them when setting improvement goals in the "Focus" step of the Recalling Experiences and Applied Learning (REAL) Model.

Some of the landmark incidents discussed here occurred in other sectors, such as aeronautics and nuclear. Learnings from these incidents is still important to our knowledge of process safety. Table 8.1 lists the landmark incidents discussed in this chapter, with their key findings.

Table 8.1 Incidents with Key Findings that Everyone Should Know

Incident	Prominent Findings and Causal Factors
Flixborough North Lincolnshire, UK, 1974	• management of change • facility siting • operational discipline
Bhopal, Madhya Pradesh, India, 1984	• culture • maintenance of barriers • stakeholder outreach • inherently safer design • chemical reactivity hazards

Table 8.1 (Continued) Incidents with Key Findings that Everyone Should Know

Incident	Prominent Findings and Causal Factors
Piper Alpha, North Sea off Aberdeen, Scotland, 1987	• culture • management of change • safe work practices • conduct of operations • Shut-down authority
Texas City, TX, USA, 2005	• facility siting (e.g., of trailers) • culture • conduct of Operations • operating Procedures • asset integrity • operational readiness • safe design
Buncefield, Hertfordshire, UK, 2005	• HIRA (insufficient layers of protection) • asset integrity • conduct of operations • vapor cloud explosions
West, TX, USA, 2013	• stakeholder outreach • facility siting • emergency management • HIRA • chemical reactivity hazards
NASA Space Shuttles, USA; Challenger 1986 and Columbia 2003	• culture • conduct of operations • HIRA
Fukushima Daiichi, Japan, 2011	• culture • preparation for natural disasters • emergency management • stakeholder outreach

8.1 Flixborough, North Lincolnshire, UK, 1974

See Appendix index entry J119

When temporary bypass piping failed, a vapor cloud explosion resulted in the deaths of 28 workers (UKDOE 1975). Many other workers suffered injuries, and significant onsite and offsite property damage occurred. The temporary piping had been installed to bypass the fifth oxidation reactor in a chain of six, which had been removed for repair.

Flixborough taught us the importance of formally, thoroughly, and consistently managing process changes, including changes that are considered temporary or emergency. The engineers at Flixborough did evaluate the changes to reactor train throughput that would be required to operate with one less reactor. However, they did not evaluate whether the planned temporary bypass piping had been properly designed to support its weight, handle the vibrations resulting from two angle bends, and endure the expected thermal stresses (Figure 8.1). After two months of exposure to stress, vibration, and fatigue, the piping failed, releasing a large cloud of flammable vapors that ignited and exploded.

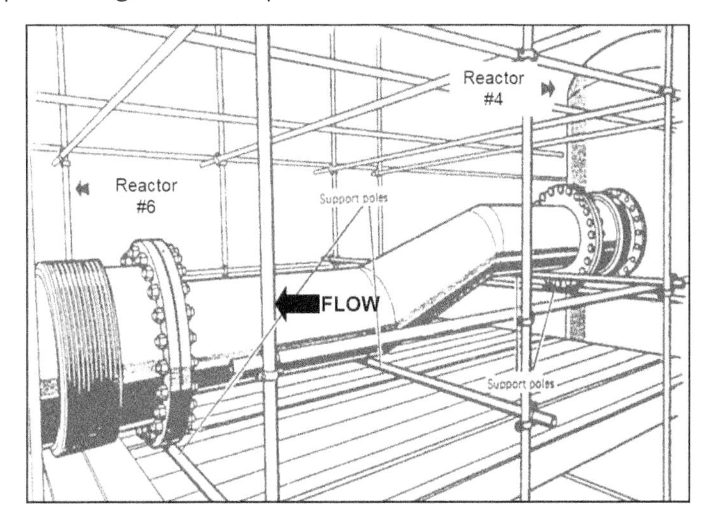

Figure 8.1 Temporary Bypass on Flixborough Reactor 5 (Source: UKDOE 1975)

Most of us stop our consideration of Flixborough there. But we can learn a lot more from this incident. The jacket of reactor 5 also failed due to a stress corrosion crack and thermal stress. The corrosion resulted from a temporary change to spray process water on the reactor head. Nitrates were added to boost heat transfer capacity. The nitrates eventually led to stress corrosion cracking of the jacket, however (Figure 8.2). Today, as then, incidents happen when companies apply temporary fixes rather than having the operational discipline to shut

Figure 8.2 Stress Corrosion Crack on Flixborough Reactor 5 (Source: UKDOE 1975)

down, conduct sound technical analysis, and perform repairs properly.

The findings from Flixborough don't end with MOC. The explosion demolished many buildings on the site. All of them were located too close to the process unit and were not blast resistant. Fortunately, the explosion occurred on a Friday night when few workers were present. Had the full weekday complement been present, fatalities could have been significantly higher.

By not stopping our investigation after we discover the primary root cause of an incident, we can identify other issues, like the need to properly space and design occupied buildings. At Flixborough, the buildings should have been designed to either keep people away from potential fires, explosions, and toxic releases, or to withstand these effects.

8.2 Bhopal, Madhya Pradesh, India, 1984

See Appendix index entry S12

Bhopal is without question the worst industrial incident ever (CSB 2014). In December 1984, water entered a storage tank containing more than 40 tons of methyl isocyanate (MIC). This resulted in an exothermic runaway reaction leading to a massive toxic gas release. The lethal vapor cloud spread over a village of tin shacks that the government had allowed to spring up just beyond the plant boundary. Unable to shelter, and too close to escape, most people in the village died or were seriously injured. In the still night air, the toxic cloud floated on to the city of Bhopal proper, affecting hundreds of thousands more people. These individuals might have been able to save themselves by sheltering in place; however, in the absence of communication from the plant about this recommended procedure, emergency responders instead told residents to flee their homes, putting them right in the middle of the toxic cloud. An estimated 30,000 people died in the weeks and months following the disaster, with countless more injured.

At the time of the incident, the safety culture at the Bhopal plant had all but collapsed. Several smaller MIC releases in the previous year highlighted the failing asset integrity system. By the date of the tragedy, nearly every barrier guarding against this tragedy had been allowed to go out of service. The chilling system lacked refrigerant. The run-down capacity was full of the MIC intermediate. The caustic in the scrubber had been used up and not recharged. The poorly maintained flare was out of service and bypassed (Figure 8.3). Plant workers disregarded pressure readings, assuming the equipment was faulty. And the uninformed action of emergency responders who told residents to flee rather than shelter in place had fatal consequences.

Figure 8.3 Bhopal Scrubber (left) and Flare (right) (Source: Dennis Hendershot, reproduced with permission)

The failures that led to the Bhopal disaster should be ingrained in our memories. Perhaps the most important takeaway is that every barrier must be maintained and functional. Furthermore, any process that requires as many barriers as those found in the Bhopal plant should prompt decision-makers to consider inherently safer design strategies, especially designs that minimize the amount of hazardous materials present. The plant exercised this strategy with its storage of phosgene, producing it on an as-needed basis and storing a minimal amount. However, it did not do the same for MIC, which is 17 times more toxic than phosgene.

Another key takeaway from Bhopal is the importance of considering all stakeholders, including the surrounding community. The public emergency response plan was nonexistent. The company did not inform local police and the public of the best actions to take in the event of an incident. On that fateful evening, the police told people who were safely sheltered in place to evacuate, sending them out into the toxic cloud.

Bhopal was a wake-up call that drove many countries to enact new laws to improve process safety and enhance emergency preparedness. It also served as a wake-up call to companies about the importance of process safety and learning from previous incidents.

8.3 Piper Alpha, North Sea off Aberdeen, Scotland, 1988

See Appendix index entry S1

Management of change also played a critical role in the destruction of the Piper Alpha drilling platform that resulted in 167 fatalities. To this day, it remains the worst offshore disaster in history in terms of fatalities. Piper Alpha was originally built to produce oil, but Occidental Petroleum later added gas production to the already crowded offshore platform. In doing so, Occidental made some questionable design choices, choosing not to upgrade the firewalls surrounding the control room to blast walls. When the explosion occurred, it caused severe damage to the control room.

MOC was not the only factor leading to the incident. The standard procedures in Piper Alpha's systems for filing permits and for communications between crew shifts were not being followed. The work permit system relied heavily on informal communication.

At the shift change, the night shift operators had learned that condensate injection pump A was out of commission for maintenance and that maintenance had not started. They were not told, however, that the pressure relief valve for pump A had been removed. It had also not been properly blanked. The work permits were not clearly displayed in the control room, and pump A was not locked and tagged out.

Further exacerbating the problem, the handover procedure from maintenance to production was dysfunctional. The procedure required maintenance and operations to meet, inspect the work site, and sign off permits together. But this did not happen. The operators were busy with shift handovers at the same time, and the practice had developed where maintenance would sign the permit and leave it in the control room or the safety office. At shift changeover, lead production operators did not review or discuss suspended permits.

When Pump B stopped operating due to a blockage, the night shift operators decided to put Pump A back in service. No information about Pump A had been communicated during the shift change, and they didn't see the permit indicating that the pressure relief valve had been removed. They restarted pump A, resulting in a gas leak and the initial explosion.

Figure 8.4 Piper Alpha Explosion and Fire in the North Sea (Source: Cullen 1990)

Making matters worse, two other platforms, Tartan and Claymore, continued to pump oil into common production line they shared with Piper Alpha. That fed the fire. The supervisor on the Claymore platform refused suggestions from a subordinate that Claymore immediately stop pumping and wasted valuable time seeking permission to do so from his superiors on shore. Tartan similarly continued to pump, amid concern about financial loss if they were to stop pumping. The oil fire ultimately melted Piper's gas pipelines, leading to a fire so hot that Piper melted to the waterline. Of the original crew of 228, only 61 survived.

8.4 Texas City, TX, USA, 2005

See Appendix index entry C11

The Texas City Refinery was the third largest in the United States at the time of this incident, where 15 contractors died and 180 employees and contractors were injured in a major explosion and fire in the plant's isomerization unit. The contractors who were killed had been working in a temporary trailer sited next to the plant's isomerization unit during a plant turnaround. As long as no adjacent units were operating, the temporary trailers could safely be sited in their convenient location; however, they should have been removed, or locked to prevent use, before the unit was restarted.

On start-up, the raffinate tower flooded and discharged to a blowdown stack. This was a design flaw—the stack was incapable of safely handling such an event. Indeed, the company standard required a closed-flare design, but that upgrade had been delayed repeatedly. The blowdown stack rapidly overfilled, creating a vapor cloud that ignited and exploded (Figure 8.5).

Figure 8.5 Texas City Refinery Explosion and Fire (Source: CSB 2007)

Most of us stop our consideration of Texas City with the siting of temporary trailers and the need for a flare system with a catch tank instead of a blowdown stack. Some recognize the culture failures that drove the decisions leading to the incident. But we can learn a lot more. For example, a proper operational readiness review—or the simple measure of walking the lines—would have revealed not only that the bottom take-off valve wasn't lined up properly, but also that the temporary trailer had not yet been removed.

The learning opportunities from Texas City don't end there. Operating procedures were not followed in part because they contained errors. Better design of control system screens would have made it obvious to operators that flow was going in but not coming out. Differential pressure-based level control for tower bottoms are inherently misleading as they can show decreasing level when the level goes above the upper sensor. In addition, workers were excessively fatigued from the turnaround. And perhaps most importantly, the fact that so many problems were allowed to persist is a clear example of poor process safety culture.

8.5 Buncefield, Hertfordshire, UK, 2005

See Appendix index entry S3

The Buncefield oil storage depot experienced the severe consequences of overflowing a large fuel tank. Fortunately, there were no fatalities. The tank had two forms of level control:

- A gauge that enabled employees to monitor the filling operation.
- An independent high-level switch (IHLS) designed to shut down operations automatically if the tank level exceeded the level setpoint.

At the time of the incident, the first gauge was stuck. Although site management and contractors were aware that the gauge stuck frequently (14 times from August to December), they did not attempt to resolve its unreliability. There was a general lack of understanding of how to use the IHLS, and as a result it was bypassed. There was no way to alert the control room staff that the tank was filling to dangerous levels. Eventually, the fuel overflowed from the top of the tank. A vapor cloud formed and ignited, causing a massive explosion and a fire that lasted five days.

Prior to Buncefield, some experts believed that vapor clouds resulting from spilled gasoline could ignite but would not explode because they are unconfined. However, an explosion did happen. This has led to many theories and a significant amount of research. Some of the factors being considered include gasoline aerosolized and well-mixed with air by cascading over the wind rib; an unusually long vapor cloud; and even trees, shrubs, and parked vehicles near the spill providing confinement (Davis 2017 and Oran 2020). This remains an area of ongoing research.

In the event of a fuel overflow, the site relied on a retaining wall around the tank and a system of drains and catchment areas to prevent liquids from being released to the environment. These containment systems were inadequately designed and maintained. Both forms of containment failed. Pollutants from fuel and firefighting liquids leaked from the bund, flowed off site, and entered the groundwater.

8.6 West, TX, USA, 2013

See Appendix index entry C74

Echoing similar trends around the world, population growth around an industrial site was an issue in the West, TX, USA, disaster. Although the facility was initially built away from the center of population, the town grew toward it, with an apartment complex, nursing home, and two public schools built between 137 and 385 meters (450 and 1263 ft) away from the West site. Zoning restrictions were inadequate considering the explosion hazard presented by the ammonium nitrate the site handled and the potential impact of an explosion (Figure 8.6).

Figure 8.6 Aerial View of West, TX, Site (Source: CSB 2016)

Ammonium nitrate is generally stable in its pure form at ambient temperatures. However, when contaminated with organic materials and other chemicals or heated, it becomes highly explosive. These properties have been exploited by terrorists to make improvised explosive devices; in the past, ammonium nitrate compositions were also used in military explosives. As a result, ammonium nitrate is regulated by US Homeland Security and by safety regulatory agencies in other countries. US OSHA and EPA regulations, however, assume that ammonium nitrate in the workplace is in its pure, ambient temperature form and it is not regulated by these agencies.

The company therefore treated ammonium nitrate no differently than most of the other products it stored, without consideration of its potential instability. The company did follow regulations for the anhydrous ammonia tanks that were located at the site. This reinforces the idea that just following regulations is not sufficient for process safety and that more information needs to be considered.

As at Bhopal, neither the public nor the first responders had been told by the company what to do in the event of an incident. Without the proper training, first responders arrived on the scene unaware of the hazards of the materials handled at the site and the procedures to handle such a fire. They made the mistake of staying and trying to put out the fire; had they been trained properly for a fire involving ammonium nitrate, they would have immediately evacuated all personnel and civilians from the facility and beyond. In all 15 people died, of which 12 were emergency responders. Hundreds more were injured in this incident.

8.7 NASA Space Shuttles Challenger, 1986, and Columbia, 2003

See Appendix index entries S9 and S10

The 1986 Space Shuttle Challenger incident is a textbook case of failed safety culture primarily driven by tremendous pressure to produce. On an unusually cold day in January 1986, the Space Shuttle Challenger broke apart and burst into flames 73 seconds after launch. The incident occurred because two O-rings in the solid rocket booster, affected by the cold, burned through, allowing exhaust gases to burn through a liquid hydrogen tank.

According to the Rogers Commission Report (Rogers 1986), the ambient air temperature at launch was 2 °C (36 °F) as measured at ground level approximately 300 meters (1,000 ft) from the launch pad—8 °C (15 °F) colder than for any previous launch. The previous evening, the National Aeronautical and Space Administration (NASA) and the solid rocket booster manufacturer had a lengthy phone call discussing the concern about launching in cold weather. The manufacturer's engineering leadership recommended a delay in the launch, but after pushback from NASA, the manufacturer's management team reversed the recommendation and provided their approval to launch.

In such a sophisticated and complex process, overruling even one no-go condition should have been unacceptable. However, after delaying the launch for a year, NASA was eager to go. NASA used the data in Figure 8.7 (top) to justify the decision to go ahead with the launch. This graph showed the number of incidents versus temperature, but it was an incomplete picture. Had NASA viewed this data along with the number of successful launches without incident, as in Figure 8.7 (bottom), they might have decided to delay the launch. The complete data showed that all launches at temperatures below 19 °C (65°F), 6 °C (10 °F) above the recommended minimum launch temperature), incurred burn incidents. In the face of pressure to produce, the poor safety culture coupled with the lack of complete understanding of risk cost the seven astronauts aboard the Challenger their lives.

Seventeen years later, NASA still had not learned the importance of maintaining a strong safety culture. During the Space Shuttle Columbia's launch on 16 January 2003, a piece of the external fuel tank's thermal insulation foam came off and struck the left wing of the shuttle. It damaged the heat-shielding panels on the left wing's leading edge. When the Columbia reentered the Earth's atmosphere, the left wing was compromised and the shuttle broke apart over Texas and Louisiana, killing all seven astronauts.

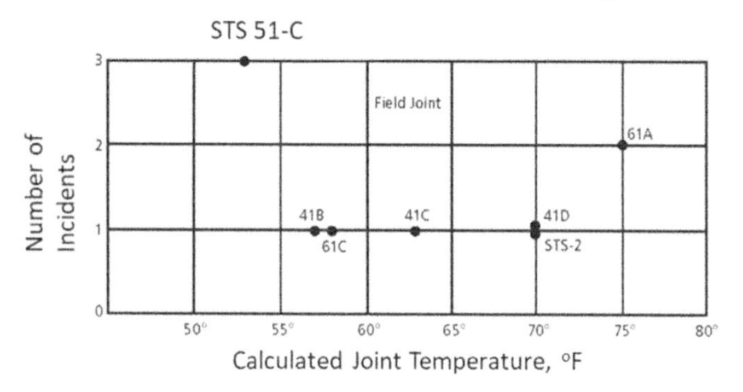

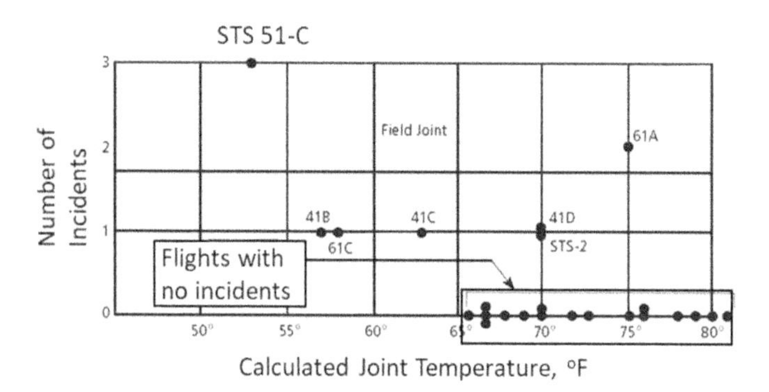

Figure 8.7 Burn-through Data from Prior Shuttle Launches (Source: Rogers 1986)

Like Challenger, this was an avoidable incident. Foam loss had been observed on almost every flight but had never done significant damage. This deviance became normalized, encouraged by a culture focused on avoiding launch delays. The Columbia Accident Investigation Board (CAIB), like the Rogers Commission, concluded that pressure to stay on an existing launch schedule was a contributing factor of the Columbia accident.

8.8 Fukushima Daiichi, Japan, 2011

According to the report issued by the Fukushima Nuclear Accident Independent Investigation Commission, the 2011 Fukushima nuclear incident was a "Disaster Made in Japan" (NDOJ 2012). The chair of the commission, Kiyashi Kurokawa, did not mince words when he stated in the report that the cause of the accident stemmed from:

> ...collective mindset of Japanese bureaucracy, by which the first duty of any individual bureaucrat is to defend the interests of his organization. Carried to an extreme, this led bureaucrats to put organizational interests ahead of their paramount duty to protect public safety.

In March 2011, the Great East Japan Earthquake triggered a severe nuclear incident at the Fukushima Daiichi Nuclear Power Plant. The earthquake damaged electricity transmission facilities supplying the plant, resulting in the need to use emergency diesel generators. Although the plant survived the initial earthquake intact, the tsunamis that followed breached its walls and destroyed both the emergency diesel generators and the seawater pumps that were critical in cooling the nuclear reactors. The tsunami also destroyed buildings and tossed around heavy machinery, making access for repairs difficult once the water receded.

The plant was built in 1967 in an area that was considered to have minimal seismic activity at the time. Its tsunami-resistant design was based on experience from the 1960 Chile tsunami. Over the next few decades, however, research had revealed an increased probability of a higher water level and velocity than observed in Chile, meaning the risk of an over-topping incident was higher than previously calculated. Engineers recommended that the floodwall be elevated and strengthened along with other countermeasures (for example, moving the backup generators up the hill, sealing the lower part of the buildings, and having some back-up for seawater pumps). However, the operating company didn't take this research seriously and postponed updating its tsunami countermeasures. Japan's regulating agency, the Nuclear & Industrial Safety Agency (NISA), continued discussing with the plant the need to update the tsunami countermeasures but did little to enforce this.

A lax safety culture could also be found in the central government, which did not have clear emergency response and evacuation plans for radiation releases. After the total loss of power occurred Fukushima Daiichi, there was concern about radiation releases. Unfortunately, communications from the central government to municipalities were slow and lacked critical information. There was confusion over evacuation procedures caused by prolonged shelter-in-place orders and voluntary evacuation orders. As a result, some residents, "were evacuated to areas with high levels of radiation and were then neglected, receiving no further evacuation orders until April," as noted in the Commission's final report (NDOJ 2012). In summary, the Commission found a lag in upgrading nuclear emergency preparedness and complex disaster countermeasures. They attributed this lag to regulators' negative attitudes toward revising and improving existing emergency plans.

8.9 Summary

Each of the 8 landmark incidents covered briefly in this chapter could justify an entire book filled with findings and improvement ideas, and for most of them, this is indeed the case. Every incident was studied deeply, and its findings shared broadly by organizations that wanted to ensure that we learn from these incidents. Considering the prominence of these incidents were and the depth in which they were studied, companies should commit to engraining their lessons in their PSMS, culture, standards, and policies to prevent similar damaging outcomes from occurring in their facilities.

In Chapter 9-14, we apply the REAL Model to scenarios typically encountered across the global process industries. We show how six hypothetical companies use findings and learnings from incidents to drive permanent change. Almost all these scenarios draw on one of the landmark incidents. While these scenarios are fictional, they are based on a range of real-world situations you might encounter. Each is intended to take the reader down a realistic path, following the REAL Model step-by-step. The fictional characters encounter all-too-common challenges that can be overcome with not only study and analysis, but also teamwork, creativity, and leadership.

8.10 References

If so indicated when each incident described in this section was introduced, the incident has been included in the Index of Publicly Evaluated Incidents, presented in the Appendix.

Other references are listed below.

8.1 CSB (2007). BP America Refinery Explosion. Report No. 2005-04-I-TX.

8.2 CSB (2016). West Fertilizer Explosion and Fire. CSB Report No. 2013-02-I-TX.

8.3 CSB (2014). Reflections on Bhopal After Thirty Years [Video].

8.4 Cullen, W. (1990), The Public Inquiry into the Piper Alpha Disaster. London: Her Majesty's Stationary Office.

8.5 Davis, S., Engel, D.M., van Wingerden, K. (2017). Large Scale Testing Confirms Deflagration to Detonation Transition (DDT) Needs to be Considered in Facility Siting, *Proceedings of the 13th Global Congress on Process Safety*, San Antonio, Texas (26–29 March 2017)

8.6 NASA (1986). Report to the President by the Presidential Commission on the Space Shuttle Challenger Accident. National Aeronautics and Space Administration.

8.7 NDOJ (2012). The Official Report of the Fukushima Nuclear Accident Independent Investigation Commission. National Diet of Japan.

8.8 Oran, E.S., Chamberlain, G., and Pekalski, A. (2020). Mechanisms and Occurrence of Detonations in Vapor Cloud Explosions. Progress in Energy and Combustion Science, 77, www.sciencedirect.com/science/article/pii/S0360128519300243 (Accessed August 2020).

8.9 United Kingdom Department of the Environment (1975). The Flixborough Disaster. London, UK: Her Majesty's Stationary Office.

9
REAL MODEL SCENARIO: CHEMICAL REACTIVITY HAZARDS

"Learn as if you were to live forever"
—Mahatma Gandhi, Indian Civil Rights Leader

The individuals and company in this chapter are completely fictional.

The Boudin Basic Chemical Company operates a large integrated manufacturing facility along the US Gulf Coast, its only site. Boudin produces a wide range of products from petrochemical feedstocks and distributes them all over the Americas.

As a small company, Boudin has only a few engineering and R&D personnel and licenses its process technology. However, the company tries to operate as professionally as larger companies, embracing Risk Based Process Safety (CCPS 2007), Responsible Care®, and a state environmental stewardship program. With limited resources, it's challenging to get everything done.

Boudin had no room to build additional capacity, and senior leadership was continuing to insist on limiting operations to a single site. All capacity expansion had to come from debottlenecking, with which the company had had great success previously. An upcoming project had been approved to replace the catalyst in four units, with the expectation that process throughput would increase by 25%.

9.1 Focus

At the annual process safety management review meeting, the chief process safety engineer, Amelia, presented the company's process safety performance to the senior leadership team. She discussed the results of a recent audit, performance metrics, and progress toward corporate goals. Members of the leadership team in turn challenged her regarding areas of slower-than-desired improvement. She managed their input deftly and got the rest of the team involved in addressing roadblocks to progress.

Overall, the presentation went well, but she paused midway through her last slide. The pause grew into seconds and became uncomfortable. Roger, the VP of operations broke the silence. "Amelia, what is it?"

Amelia explained that she felt uneasy about the company's effort to upgrade catalysts in four units on the site as part of a general capacity expansion. She acknowledged that the catalyst vendors had assured Boudin engineers that the new catalysts were completely compatible. While conducting an MOC review for the project, however, she found that the vendors had not provided any data to confirm this. "It might be nothing," she said, "but I feel extremely uncomfortable if we move forward without having done the professional analysis."

After some discussion, the VP asked Amelia for her recommendation. She suggested a combination of literature search and lab testing and provided a budget and schedule for these actions. After some discussion, the leadership approved her plan.

9.2 Seek Learnings

Amelia searched Boudin's internal incident and near-miss reports to see if there had been any reactive chemical issues but found none. She then assigned the task of reviewing the literature to Jason, a senior engineer. As Jason expected, he didn't find any information about the new catalyst in the chemistry literature. He turned to publicly investigated incidents that might be applicable and found one incident report that closely matched Boudin's situation:

Moerdijk, Netherlands, 2014

See Appendix index entry D9

Two explosions rocked a petrochemical plant following a runaway reaction that generated pressure that could not be relieved. Two workers died in the explosion.

The plant converted ethylbenzene to styrene monomer, producing propylene oxide as a byproduct. The explosion occurred in a reactor where intermediate methylphenylketone (MPK) was hydrogenated to methylphenylcarbinol (MPC), using a proprietary catalyst. This was the first start-up following the replacement of spent catalysts with an improved catalyst that the plant had not previously used.

Before introducing MPK, the catalyst bed was heated by circulating ethylbenzene through the reactor and an external heat exchanger. The company did not know that the new catalyst could react with ethylbenzene at temperatures that could typically be reached during start-up due to normal temperature fluctuations.

Due to the reactor fluctuations, the liquid level in the vapor-liquid separation tank on the flare line also fluctuated widely. Every time the tank reached high level, an interlock would close the valve between the tank and the flare and had to be manually re-opened by an operator when the liquid level dropped. Just before the runaway, an operator had neglected to re-open the valve to the flare.

A thermal runaway occurred. With the valve to the flare left closed, the resulting pressure could not be relieved, leading to the explosion.

Although Jason was not able to find other public cases of runaway reactions involving catalyst pellets, he did find several other cases where unexpected runaway reactions occurred:

Oita, Japan, 1996

See Appendix index entry J30

During a trial batch of a new pesticide, an intermediate was held at high temperature during a process delay due to a pump failure. The intermediate self-reacted exothermically, leading to an explosion that injured an operator and damaged the production building.

Kitakyushu, Japan, 1996

See Appendix index entry J35

A contaminant present when raw material was fed to a resin intermediate process led to an explosive decomposition of the raw material. The contaminant back flowed from the vapor treatment system. While no one was injured, the plant was destroyed and did not restart. The contaminant had entered the vapor treatment system from another part of the process.

Lodi, NJ, USA, 1995

See Appendix index entry S11

Water leakage into a blender led to an explosion that resulted in five fatalities and injured four. The explosion destroyed the plant and nearby businesses and forced 300 nearby residents and students to evacuate their homes and a school. The solids were known to be water reactive, but the blender had a water-cooled seal that may have leaked. An additional source of water may have been the liquid feed line to the blender.

Bhopal, India, 1984

See Appendix index entry S12

Contamination with water and rust in a tank of methylisocyanate led to a runaway reaction. With all mitigative barriers disabled, toxic gas spread over the city, resulting in the deaths of more than 30,000 of the city's residents.

9.3 Understand

After studying the above reports, Jason came to understand that the following findings had the potential to apply directly to Boudin Petrochemicals' catalyst project:

- *Moerdijk*: A material used only for start-up reacted with the new catalyst under conditions that could be expected to occur during start-up.
- *Oita*: An intermediate was stable enough to be held a short time in process, but during a long hold, the decomposition rate accelerated.
- *Kitakyushu*: A process material back flowed into the reactor, catalyzing a runaway reaction.
- *Lodi and Bhopal*: Materials normally used for nonroutine activities contaminated the process, causing a runaway reaction with process materials.
- *Bhopal*: Corrosion byproducts served as a catalyst, accelerating the runaway reaction.
- *Oita, Kitakyushu, and Lodi*: Incidents occurred outside of normal, expected operating conditions.

Jason also noted some findings for future consideration:

- *Moerdijk*: The reactor could not be vented safely because the vent valve required operator action to reopen after it interlocked closed.
- *Bhopal*: Many barriers that could have prevented or mitigated the consequences of the disaster were in place but out of service.

9.4 Drilldown

Having read and digested these publicly reported incidents, Jason looked for "hidden" findings—findings that were either not emphasized or not discussed in the reports or that seemed to be open questions. He made notes for each report:

- *Moerdijk*: The reasons for reactor instability during the heat-up step did not appear to have been understood, and the potential problems resulting from that instability didn't appear to have been considered. Although the unstable conditions before the incident seemed to have been expected, were they worse than normal? Could the catalyst change have made the instability worse?
- *Oita*: This incident is an interesting parallel to Moerdijk. The incident occurred not during production, but during a time the process was hot but not running. In both cases, the companies did not appear to have considered hazards when the process was in a static phase (i.e., on hold or recycle).
- *Bhopal*: Corrosion products could have resulted either from choosing the wrong construction materials during design or from substituting the wrong materials during construction or maintenance.
- *Kitakyushu and Lodi*: Contaminants appear to have resulted from engineering design errors: backflow prevention in the case of Kitakyushu and failure to disconnect water sources in the case of Lodi.

Jason noted additional "hidden" findings that applied generally:

- *Bhopal*: For a plant to operate when nearly all barriers are out of service requires a total breakdown of the culture. Senior leadership may have been well-meaning but were detached from the operation. Plant leadership allowed deviation from required maintenance as well as operation with barriers out of service.

- *Moerdijk*: When operators are struggling to start up and gain control of the process, it is easy to be distracted. In that environment, it isn't too surprising that the operator failed to re-open the vent valve after it had interlocked closed.

- *Oita, Kitakyushu, and Lodi*: The conditions under which these runaway reactions occurred could reasonably be expected to happen multiple times during the lifecycle of the processes. As such, these conditions should have been addressed in operating procedures. The procedure should have specified that operators must verify that a vessel is clean and empty before starting it and that hold/idle conditions are safe.

9.5 Internalize

Jason met with two colleagues from engineering and production, Emma and Phillip. They discussed the core and deeper learnings that Jason had extracted from the five external incidents and compared the findings with the planned catalyst changes. They examined the vendor data addressing the kinetics and byproduct spectrum for the four processes at a range of temperatures above and below normal operating conditions.

They then examined process historian data from the last startups of the units. In each case, temperature fluctuations that went above the range covered by the vendor had occurred. Emma suggested that they could add a safety instrumented system (SIS) to ensure the temperature stayed within the known range of the vendor data. Phillip objected, pointing out that if the SISs kept shutting down the units, they'd never get started up. They agreed to defer seeking solutions until they got more data.

Emma then pointed out that, although none of the processes heated up with a non-process chemical, as in the Moerdijk incident, they did heat up with just one of the reactants. Vendor data didn't address that scenario. She also noted that when the reactor went on hold, operations would be stable once the remaining reactants converted to the final product.

The three agreed that each of the four catalysts needed to be tested in an external reactive screening laboratory, both with the startup reactants and with the final product. Testing with the reaction mixture wasn't necessary, they reasoned, because in the presence of catalyst the mixture would be converted to final product. They considered whether to test reactivity in the presence of contaminants but concluded that there was no mechanism for contaminants or corrosion products to enter the process.

When the results came back, two of the four processes had self-accelerating decomposition temperatures (SADT) for the startup reactants in the presence of the new catalyst that were less than 50°C above the maximum temperature typically seen at start up. For these same processes, the SADTs were 75°C above the stable hold temperature with final product, but the time to maximum rate (TMR) for final product was less than 24 hours.

The three were about to jump to an engineering solution, but Jason pulled them back to discuss the situation more strategically. He reminded his colleagues what they'd just learned: Several of their processes had potential chemical reactivity hazards that they had not recognized before. He recommended that they first develop a design standard addressing reactivity hazards, including when and how to test, and how to design and operate the process, depending on the test results. This standard would not only guide their current efforts but also inform future process designs and modifications.

9.6 Prepare

Jason, Emma, and Phillip invited Chip, the chief engineer, to join their group to draft the chemical reactivity hazard management standard and plan the path forward. They also invited Maria, an instrument and controls engineer, who was assigned the task of developing the necessary interlocks and logic to meet the eventual standard.

They drafted the standard with input from the testing lab and from CCPS literature (CCPS 1995), then circulated it to key stakeholders in engineering, production, and HSE. About half of the reviewers resisted the standard, citing the cost and claiming the proposed temperature and time limits were too conservative and rigid.

Emma responded by providing time-temperature graphs for all the holds the site had experienced over the past three years. She overlaid the TMR test graph for the appropriate new catalyst. From this presentation, it was clear that the site had had five holds during that period where the plant was on the brink of thermal runaway. The critics then saw the wisdom of the recommended conditions. The team noted that they should have done the overlay analysis as a first step.

With agreement from the stakeholders, Maria provided the draft standard to the technology licensor and began discussing with them the hardware and software that would be needed to control temperature within the range specified in the draft standard.

The team then developed the communication and training associated with the standard. They agreed that in addition to posters which displayed the hold time and temperature limits for the two processes, they would develop a simple simulation to show how fast the reaction would run away if the time and temperature limits were surpassed. They also wrote a report summarizing their search of external incidents, the testing, and the recommendations they derived, to include in the process safety knowledgebase.

9.7 Implement

Amelia brought Jason and his colleagues to the leadership team to present their recommendations. They showed the overlay graphs for past hold scenarios and similar graphs for high temperature excursions during normal operations. This got everyone's attention, and they received no pushback on the need for the standard or the investment in control equipment.

Already impressed with the analysis, leadership team members debated the communication plan among themselves, adding a few suggestions for improvement. After the proposal had been addressed to everyone's satisfaction, Roger, the VP of operations leaned forward and said, "Please proceed with your plan. Is there anything specific you need us to do?"

Amelia replied, "When we've completed the turnaround and are ready to start up with the new catalyst, we'd like two or three members of the leadership team to participate in the awareness campaign around the new standard and the implications in the new design."

Enthusiastic nods around the table signaled the leadership team's agreement. "One more thing," Amelia added. She brought up a new PowerPoint slide summarizing the general findings Jason had extracted from his review of publicly investigated incidents. It read:

Recommendations

- Add standing agenda item: Review status of all barriers.
  - Why any barriers are out of service.
  - Any common reasons for barriers being out of service.
- Initiate special program: Identify error-prone situations.
  - Review progress towards eliminating them.

Amelia said, "We generally do a good job maintaining our barriers. But when we re-studied the Bhopal disaster, it became clear that we as a leadership team need to make barrier management a permanent agenda item. One area of barrier management we must improve is where we expect our operators to take action. From just a casual walk-through, we found dozens of situations where operators could find themselves distracted or otherwise unable to perform the functions we need them to do. I'd like us to address this once we're through the turnaround."

9.8 Embed and Refresh

Three years later, Jason was now the chief process safety engineer. He'd been promoted six months earlier following Amelia's retirement. Along with Amelia, all but one member of the leadership team had also retired. The new environment was less formal, and most days, Jason enjoyed the work.

> **10 or 10 Means Shut Down**
>
> 10-hour hold, 10° overtemperature
>
> 6 runaway reactions prevented since 2014

Today, though, Jason was worried. As he entered the plant, he paused by the prominent sign that read, "10 or 10 means shut down." He'd created the sign three years ago to raise awareness of the new reactive hazard standard. Since then, the reactors had tripped six times. After the first one, Amelia convinced him to add the bottom line to the sign, to reinforce to workers as they entered the plant what could happen if they didn't shut down when they should.

But maybe the reminder wasn't working anymore. Jason had gotten a call at 4 a.m. asking him to authorize extending a hold until 10 a.m. One feed pump was being rebuilt in the shop and the other had failed. They were rushing the repair of the first pump and had ordered parts for the second. One pump or the other would be ready to run by 10 a.m. at the latest, the caller told him.

"Can't," Jason had managed to say, trying to clear the sleep from his head.

"Oh, right," the supervisor replied. "There's some kind of standard."

Still standing in front of the sign, Jason sent a quick email to his assistant, asking him to change the number of disasters prevented from six to seven. Then he added a reminder to his calendar to have the VP of operations visit the unit to offer an "Attaboy" to the workers and thank them for shutting down, and then follow up with a message to all personnel. As he stood there, Phillip stopped next to him. Phillip's promotion to VP of operations had happened about the same time as Jason's promotion. "It looks like you're about to send me an email," Phillip said. "I heard about the call you got. Stop by my office at two today and we'll talk about it."

When Jason arrived at 2 p.m., the marketing communications director, Sarah, was in Phillip's office. Phillip explained that he had asked Sarah to help refresh the messaging around the reactivity standard. Sarah explained that people get used to seeing signs and eventually don't see them. She showed them a communication plan she'd developed to deliver the message in different ways at intervals over the next three years. The plan factored in the different ways that individuals learn. It included videos of related incidents that shift teams and other work groups could play at safety meetings, as well as a simulation, a sign contest, a new e-learning module, and a workshop. At the end of the three-year period, they would repeat the plan.

Remembering the strategy his predecessor Roger had used, Phillip asked each of his leadership team members to take responsibility for a part of the communication plan. He personally volunteered to organize the sign contest. Judging took place at the plant's pre-Mardi Gras shrimp and crawfish boil. Phillip held up the signs submitted by 12 teams, one at a time, for everyone to see and offered positive comments on each one. After careful deliberation, he declared every sign a winner, and said that the plant would use them in rotation, one per month. He closed by announcing the new e-learning module, workshop, simulation, and videos.

But he left the "10 or 10" sign by the front gate. Over the next three years, although the number of disasters prevented continued to rise, he never had to change the "since" year and reset the number to 0.

9.9 References

If so indicated when each incident described in this section was introduced, the incident has been included in the Index of Publicly Evaluated Incidents, presented in the Appendix.

Other references are listed below.

9.1 CCPS (1995). *Guidelines for Chemical Reactivity Evaluation and Application to Process Design*. New York: AIChE.

10
REAL MODEL SCENARIO: LEAKING HOSES AND UNEXPECTED IMPACTS OF CHANGE

"A [person] who carries a cat by the tail learns something [they] can learn in no other way."—Mark Twain, Author and Humorist

The individuals and company in this chapter are completely fictional.

Feijoada Pharma produces a range of generic drugs in a modern facility outside São Paulo, Brazil. Each of its five primary reactor trains run one- to three-month campaigns to produce active pharmaceutical intermediates. The reactors are then reconfigured for the next product.

Each reactor can be fed via multiple routes, including:

- dip pipe
- free-fall from a nozzle in the reactor head
- free fall via one or more spray balls in the reactor head
- through a recirculation line.

These feed routes are accessed through four hose connections near the reactor that are grouped close together for convenience but are clearly labeled. The materials to be fed are similarly piped close to the reactor, terminating in hose connections. These, too, are grouped close together for convenience and clearly labeled.

Operators connect the desired feed material to the desired feed port using the appropriate flexible hose designated in the operating procedure. Raw materials can also be fed to the reactor through any of the ports from drums. Some hoses are kept connected for the duration of the campaign, while others are purged and disconnected to allow a different raw material to be fed via a given feed route.

Feijoada receives raw materials in tank truck quantities in its tank farm. A single unloading pad serves the plant. The driver must connect to the correct hose connection to unload the material into the intended storage tank. These hose connections are also clearly labeled, and the connections, grounding, and bonding are verified by an operator.

10.1 Focus

Ana Maria, the manufacturing leader, entered the conference room and greeted her leadership team, seated around the room. She dropped a thick manila folder on the table and asked, "What's going on with our hoses?"

João, the plant superintendent, replied, "We've had a lot of leakers, lately. Juliana and I have investigated this. Basically, we bought them all at the same time and they're wearing out at the same time."

Juliana, the maintenance manager, shook her head. "It's true that we bought them at the same time and they're failing at the same time," she said, "but my friend João is oversimplifying the issue." A tense look passed between them before she continued, "I wouldn't expect every hose to have the same life. Every process and every chemical are different."

Antônio, the process safety leader, spoke up. "We really need to be focusing on the potassium cyanide leak last week," he said. "We caught it on the detector right away, so we didn't lose more than a few drops. No one was exposed. But it could have been much worse."

Everyone around the table began talking at once. After a minute, Ana Maria rose from the head of the table, causing the room to go quiet again. "We need to understand why all our hoses are failing at the same time. I don't know if it's in how we operate, how we maintain, how our process was designed, or if all the hoses came with a manufacturing defect. It doesn't matter. Let's find out why and fix it before the next leak is a disaster."

10.2 Seek Learnings

Francisco, the engineering manager, joined João, Juliana, and Antônio for lunch at the cafeteria. After some small talk, they began discussing a plan for addressing the issue with hoses. Francisco agreed to check material compatibility. João would check that hoses were being used properly. Juliana made a note to verify the inspection intervals and to test the hoses that hadn't leaked yet.

Antônio agreed to check the public literature for similar cases and anything else that might inform their path forward. He quickly identified several cases involving hose leaks and failures:

Kawasaki, Kanagawa, Japan, 1989

See Appendix index entry J57

A hose used to connect a fuel oil tank to a boiler broke, leaking oil into the sea. The hose was attached to the tank and boiler, hanging freely with no support. The weight of the filled hose exceeded the tensile strength of the hose, leading to failure.

Festus, MO, USA, 2002

See Appendix index entry C23

A braided metal hose being used to unload chlorine from a rail car ruptured, releasing more than 20,000 kg of chlorine. Three workers and 63 nearby residents sought medical treatment. The hose was stamped Hastelloy C, the intended metallurgy. However, the hose was found to be stainless steel.

Belle, WV, USA, 2010

See Appendix index entry C25

A short segment of small diameter PTFE-lined braided stainless-steel hose was being used to transfer phosgene from cylinders to the process. The hose burst, spraying a worker with phosgene. The worker died several hours later. The phosgene hoses were supposed to be replaced monthly. However, the hoses had not been replaced in nearly six months. Further investigation revealed that when the asset integrity management system was updated, the replacement notification had been inadvertently cancelled.

Atchison, KS, USA, 2018

See Appendix index entry C49

Raw material sulfuric acid was being unloaded by a supplier into a customer's plant. The driver attached the hose to the wrong connection and pumped sulfuric acid into the sodium hypochlorite tank. This created a large release cloud of chlorine gas from the tank vent. As a result of exposure, 120 employees and nearby residents sought medical attention.

Antônio also recalled reading about hose failures in several issues of the *Process Safety Beacon* (CCPS 2020). He found the following cases in the newsletter archive:

Anonymous 1

Process Safety Beacon July 2007

Offshore oil platform workers were unloading a tote of methanol into a storage tank using a flexible hose. When the tote was lifted to drain methanol into the storage tank, methanol sprayed from the side of the hose and caught fire. One worker was burned fighting the fire. Investigators found that the hose had previously been split. Instead of discarding the hose, a worker tried to repair the split with duct tape.

Swansea, SC, USA, 2009

Process Safety Beacon Nov. 2015

A truck of anhydrous ammonia was being unloaded when the unloading hose failed catastrophically. Ammonia spread over the highway, exposing one motorist. The nearby community sheltered in place. The hose used to unload ammonia was labeled "For LPG unloading only."

Anonymous 2

Process Safety Beacon Mar. 2009

A driver was directed by an operator to connect the hose for a shipment of sodium hydrosulfide to the wrong connection. This material was rarely delivered, and the operator thought the truck contained the more routinely delivered material ferrous sulfate. The two chemicals reacted, forming a toxic cloud of hydrogen sulfide that was fatal to the driver.

Anonymous 3

Process Safety Beacon Dec. 2008

A tote was being filled with flammable ethyl acetate using a flexible hose. The operator heard a "pop" and found the tote engulfed in flames. Investigators found that the hose and tote were not properly grounded and bonded.

Before reporting his results, Antônio thought it would be a good idea to go beyond the list of hose leaks in Ana Maria's folder. He reviewed other work order records related to hoses and piping that were unrelated to leaks. He found nothing obviously related to hose leaks, but he was curious about the recent spike in work orders to repair pipe hangers and other piping supports. This finding didn't apply to hoses, but he made a note to bring it up to Juliana as another issue they might need to address.

10.3 Understand

Antônio summarized the findings that might be relevant to the hose leakage problem:

- *Kawasaki*: Excessive stresses on the hose.
- *Festus and Swansea*: Incompatibility of hose material with process materials
- *Belle and Anonymous 1*: Hoses used well beyond their service life. This represented an asset integrity failure.
- *Atchison and Anonymous 2*: In both cases, hoses were connected to the wrong fitting. (Although the findings from one of the cases did not seem to directly apply to Feijoada's problem, Antônio noted it for future consideration.)
- *Anonymous 3*: The case resulted from improper bonding and grounding.

10.4 Drilldown

Digging deeper, Antônio considered how the failures in all the cases came about. In the case of the CSB-investigated incidents, the root causes were clearly spelled out. However, he had to use his imagination when considering the other cases.

In the case of Kawasaki, he assumed that the tank connection and boiler feed connection were both near grade level. If the hose was too short, workers might have been able to stretch it taut between the two connections, lifting the hose off the ground. The weight of the hose's contents would be applied at an angle to the hose fitting, multiplying the force exerted. Additionally, a hose hanging about a meter above grade could be subject to impact from foot and vehicular traffic. If the hose had been strung up above a roadway between two posts, however, it could have swayed in the wind, causing it to rub at the points of contact with the posts. Moreover, why was the Kawasaki plant using a hose when a hard-piped connection seemed more appropriate for an application like this?

Either way, the Kawasaki plant did not seem to have completed a thorough hazard evaluation, running the hose without considering the possible impact of foot traffic, stresses, or rubbing on sharp corners.

Festus and Swansea clearly had two different causes. In Festus, the vendor mislabeled the hose and the plant took the vendor's word for it. In Swansea, the label was correct, but the driver ignored the label. João was already checking whether the correct hoses were being used, but could hoses be coming in mislabeled? He recalled that they'd changed vendors in the past few years. Could the new vendors' hoses have different colors than the old ones, creating confusion?

In the cases of Belle and Anonymous 1, problems clearly existed in both plants' asset integrity management systems. Hoses used well beyond their service life represented an asset integrity failure. Juliana was already looking into their own testing and inspection frequencies.

But the use of duct tape in Anonymous 1 worried him. Could the failures of some hoses have been delayed by temporary (and forbidden) patching? And only reported to management grouped in with later failures? The operators would never do this, would they? It would be easy enough to check. Just walk into the control room and ask innocently, "Can I borrow your duct tape, friends?" They had no reason to keep duct tape on hand.

At first, Atchison and Anonymous 2 didn't seem to apply to the current problem, although Antônio made a note to do a human factors evaluation of the storage tank unloading system. However, when he thought about it more deeply, connecting a hose to the wrong reactor port could cause unexpected chemicals to get into the hose. He made a note to follow up on that possibility.

One of the other incidents in the Belle report, the oleum leak, caught Antônio's attention. The piping was made of the correct material for transferring oleum and could certainly resist the moisture in ambient air and steam. However, trace oleum vapors had escaped through a flange and mixed with moisture in the air, forming sulfuric acid of less than 98% concentration. That composition was very corrosive to oleum piping! Could the failing hoses be resistant to each individual material they carried but incompatible with mixtures? The oleum piping failure was happening from the outside-in. Could their hoses be failing for a similar reason?

Antônio didn't think Anonymous 3 applied to their situation. The plant instrument techs were religious about inspecting and testing grounds, cables, and continuity. But more importantly, none of the hose failures appeared to be caused by burning, even localized burning. Perhaps hoses rubbed across cables, though. He'd have to check that.

10.5 Internalize

A week later, Antônio met again with João, Juliana, and Francisco to compare what they'd learned. Francisco reported that in every case, the proper hoses had been specified. João reported, with obvious relief, that operators were double verifying each time they made a hose connection and no exceptions had been found.

Juliana shared the inspection and maintenance records. The inspection and replacement intervals hadn't changed in five years. In the past year, 99.5% of hose inspections and tests were completed on time. While not perfect, the delay for the other 0.5% was far too short to explain the observed problem. Antônio reported that he found no duct tape in the control room and locker room, and that in fact the plant had not purchased duct tape in several years.

Antonio also shared what he learned from the external incident review, and the four colleagues developed a plan for further study. They met again a week later to discuss their findings, summarized in Table 10.1.

Table 10.1 Results of Study Based on External Incidents

Study plan action	Feijoada plant finding
Evaluate how hoses might be subjected to rubbing, snagging, and impact.	None found in the process of connecting, using, and disconnecting hoses.
Positively identify hose material of construction.	All materials of construction verified correct.
Check if new vendors' hoses look the same as older hoses.	The new hoses looked slightly different but were marked more clearly than the old ones. Operators felt the new hoses presented a smaller chance of mix-up.

Table 10.1 Results of Study Based on External Incidents (Continued)

Study plan action	Feijoada plant finding
Determine potential for multiple materials to mix inside hoses. If found, test resistance to mixtures.	After careful review of procedures for flushing hoses and compliance with the procedures, opportunities for mixing in hoses were not found.
Determine if leaks or drips might be impacting hoses.	None found.
Evaluate where flexible hoses could be replaced with hard-pipe connections.	Several opportunities were found, but none of the hoses in these services were leakers.

"Maybe we just haven't been asking the right questions," Antônio said. "Let's take one more walk up to the production floor." As the four colleagues walked into the building, Juliana turned to João. "The lighting in here is so much brighter now since you installed the LED bulbs," she said. "My mechanics really appreciate how much better they can see what they're doing."

Just then, Antônio noticed Adriana, a new operator, standing on water feed piping to read a gauge. He opened his notebook to the page where he'd noted the issue about pipe hangers and showed it to his three colleagues. They walked over to Adriana. "Are you having trouble reading that gauge?" Antônio asked her.

"Yes," she replied. "Ever since the new lights were installed, we get a lot of glare off the glass. We have to get closer to read a lot of gauges."

"Does that include the gauges over the raw material manifolds?"

She nodded. "Are you going to tell me we shouldn't be standing on the hoses to read those gauges?"

"Or the pipes," João agreed.

10.6 Prepare

On their way back in the conference room, João said, "Why don't we have the operators figure out the best way to read the gauges without standing on the pipes or hoses?"

"That makes sense," said Juliana. "I'll ask my lead mechanic Paulo to coordinate to make sure maintenance is not affected.

"I like it," Antônio said. "And we can focus on developing policy, training, and communication."

A few days later, the group walked through the plant with Adriana, the operator. She pointed to the gauge they'd seen her reading a few days earlier. "The guys were arguing whether to make hoods, like on traffic lights, to shield the gauges, or whether to buy the hoods from the traffic light manufacturer. But I just bent the support bracket a little bit to tilt the gauge, and the glare is gone. It shouldn't take more than a week to get them all adapted. We're going to need maintenance to make us up a few special brackets."

At the next leadership team meeting, Antônio summarized their plan:

- *Adjust the gauges* to eliminate glare (Lead = Adriana).
- *Draft a policy* forbidding standing on hoses, piping, or equipment (Lead = Antônio).
- *Add the policy requirement* to standard operator, mechanic, and contractor training (Lead = João).
- *Develop a communication plan*, working with the public relations group (Lead = Antônio).

Juliana added, "We also identified three hoses that can be replaced with hard-piping. We recommend making those replacements also."

Ana Maria complimented the team on considering inherently safer design and on finding a solution with almost no cost. She assigned Márcia, a communications specialist, to work with Antônio to develop posters and to modify the site briefing video.

10.7 Implement

The gauges were adjusted within a week as Adriana promised. Antônio decided that instead of writing a new policy, he would simply modify the existing "Work at heights" policy, adding two sentences:

> *"Personnel must never stand or climb on piping, hoses, conduit, or equipment. To access elevated equipment and instruments, use only the means specified in this policy for working-at-heights."*

After discussing the contents of the modified policy with Antônio, Márcia developed a poster showing a monkey climbing, with a big red X through it, with text explaining that the only appropriate places to stand were the floor, a ladder, a scaffold, or a person-lift. The posters were distributed around the plant. An image of the poster was also incorporated into the plant training video.

Don't be a monkey!

Stand only on:

- the floor
- a ladder
- a scaffold
- a person-lift.

10.8 Embed and Refresh

About 18 months later, Antônio pulled into a carpark spot next to Adriana. After they'd wished each other "*Bom dia*," Adriana said, "I've been meaning to ask you, Antônio. You addressed the problem of stepping on hoses by changing the work-at-heights standard. And I guess that was a good thing because it also protects insulation, piping, and cables. But it really doesn't prohibit us from stepping on a hose that's on the floor."

They passed through the gatehouse, exchanging hugs with the guard. As they stepped back outside, Antônio smiled at her. Adriana was becoming a seasoned operator. Someday she'd take João's job for sure. They stopped in front of the monkey poster. He pointed at it and said, "Doesn't say you can stand on a hose, does it?"

"So, walking on it should be OK!" she exclaimed, and slipped through the locker room door.

Antônio got a double coffee from the cafeteria and sat alone at the back, thinking about Adriana's joke. It was a joke, wasn't it? He thought about the new reactor system that would be added into the high bay. They would be hiring new operators and running more complicated processes. There would be a detailed HAZOP, ongoing construction activities, and new operations. Lots of opportunities for things to fall through the cracks.

He reflected on the work they'd done to figure out the reason the hoses were failing. Initially, they'd checked everything that anyone could expect to be a problem and found nothing. Who would have expected that a change in lighting would cause operators to deviate from their normally diligent performance?

He started a new email to himself and typed *There is always something else you didn't consider*. For example, he knew he'd have to address walking on hoses. And what about traffic or dragging sacks across hoses? It would be easy enough to update the training, and for that matter to draft a specific hose policy.

But that wouldn't be enough. Yes, a different campaign would be needed. He added to his email: *What else did we miss?* That would be a good name for the campaign. It could be a contest. Name any hazard in the plant that didn't seem to be recognized. Prizes could be given for the best individual response, the best team response, even the best response from someone in management! The best poster, the best video, maybe even the best skit or song. He was sure there were more prize opportunities.

The campaign would not only remind all employees of current known hazards and the barriers, standards, and policies to protect against them, it might also reveal something else they had missed.

He sent an email to Ana Maria, asking for a meeting the next day to pitch the idea. Draining the last of his coffee, he rose and headed to his office, already drafting the plan in his head.

10.9 References

If so indicated when each incident described in this section was introduced, the incident has been included in the Index of Publicly Evaluated Incidents, presented in the Appendix.

Other references are listed below.

10.1 CCPS (2020). *Process Safety Beacon*. www.aiche.org/ccps/ process-safety-beacon (accessed June 2020).

11
REAL MODEL SCENARIO: CULTURE REGRESSION

"Knowing is not enough; We must apply. Willing is not enough; We must do."—Bruce Lee, Martial Artist, Entertainer, and Leader

The individuals and company in this chapter are completely fictional.

The Lamington Oil Company operates in the East Timor Sea, about 500 km off the coast of northern Australia. The company's offshore rig was built in 2005 in the rush to tap into the speculated USD 50 billion worth of oil and gas in the fields.

The rig operates 24 hours a day, seven days a week. The employees on the rig worked 12-hour shifts with two short breaks plus a lunch break. They worked on a demanding 14/14 rotation (14 days on the rig followed by 14 days off the rig), but fatigue risk assessments were not carried out regularly.

At the beginning, the rig crew was meticulous about following all written procedures. No short cuts were taken. Equipment maintenance was done as scheduled and all the needed work permits were filed properly. The original rig manager made personal and operational safety top priorities. He made sure that every shift started with a safety meeting where everyone was engaged. But when the rig manager left for an onshore position, his successor had a different perspective. The new rig manager said he was concerned about safety, but it was clear to the crew that he put production numbers as the top priority. Maximum production always was the new goal he constantly stressed to everyone on the rig.

The new rig manager's message may have had some unintended consequences. The rig's safety officer saw an uptick in minor accidents such as slips and falls. But he didn't feel compelled to report them. He wanted the safety record of the rig to remain stellar. He just thought he was doing what the boss wanted. The rig's production and maintenance supervisor were under a lot of pressure to meet his numbers. He began to stray from the equipment maintenance schedule and doesn't have as much time to make sure that all the proper permits are filed when there is maintenance work.

11.1 Focus

Charlotte was a recent chemical engineering graduate who just started her new job as a chemical process engineer on the offshore rig. Charlotte had always been adventurous and not one to shy away from challenges. She knew that working on a male-dominated rig had its problems, but this was her dream job. Charlotte was also vocal and used to sticking up for herself and her beliefs. One of her responsibilities was conducting risk assessments under the guidance of her boss, Lucas, the rig's safety officer.

Charlotte walked into Lucas's office and said, "We've talked about conducting risk assessments, but they have primarily been focused on equipment failures. How would you feel about conducting one on human fatigue?" Lucas looked at her warily. He was pleased with her performance so far, but he felt she was always pushing him to do more.

He said, "And what do you want to get out of this?" Charlotte replied, "The rotation is pretty tough on the crew, and from what I learned in university, human factors is often overlooked when it comes to safety. Fatigue results in slow reactions, reduced ability to process information, decreased awareness. Bottom line is that it can result in accidents."

Lucas thought to himself, "She has a point, and maybe this will help me figure out the cause of the recent mistakes a couple of the crew have made." Lucas told Charlotte that first, he would have to check with Oliver, the production and maintenance supervisor. After all, Charlotte would need to conduct some personnel interviews with Oliver's crew. Plus, he needed to get the approval of the rig manager, Mason.

Lucas brought up the request with Oliver, who eventually agreed, but not without some pushback. "My crew already has a lot on their plates," Oliver said. "Taking time out for interviews is just another thing for them to do." Lucas replied, "We're all under pressure. But think of it this way. Fatigue has been a common factor in major accidents. Last thing we want on our hands is something big to happen on the rig. Let's make a compromise. Instead of interviews, let's do it as a survey. We'll set it up so that it doesn't take more than fifteen minutes to complete." Oliver said, "Fine, but I'll hold you to it. No more than fifteen minutes."

Lucas next set up a meeting with Mason to discuss the survey. He was able to win Mason over with his focus on the business case for process safety. Like Oliver, Mason told Lucas that the survey should not impact operational efficiency.

Lucas told Charlotte of the compromise, and she quickly went to work setting up a survey. She thought to herself, "While I'm at it, I'll throw in a couple of questions to see what people think of the process safety culture on this rig." A few weeks later, the results were in. Charlotte analyzed the results and immediately took them to Lucas. She said, "Looks like we have a problem here. Fatigue is definitely an issue and perhaps even more problematic is the process safety culture. Seems like folks are willing to cut corners to meet production numbers."

Lucas was distressed by the results. He thought of his chats with Oliver and Mason, when he had reminded them that fatigue was often a cause of major accidents. Poor process safety culture falls into that category too. He told Charlotte, "We have to make a case for improving our process safety culture before we meet with Mason. Let's start by looking for external incidents and see what we can learn."

Lucas continued, "Charlotte, you're probably too young to remember the Space Shuttle disasters. But let's put it this way, there was so much pressure to get the shuttle off the ground that the management ignored the engineers' warnings. We can't let that happen to us on this rig." Charlotte nodded in agreement. She may not have been born when the Challenger exploded, but she remembered her parents talking about the incident when she was growing up.

11.2 Seek Learnings

Charlotte quickly found several relevant major offshore incidents, including some that were in their region:

Piper Alpha, North Sea, UK, 1988

See Appendix index entry S1

Communications broke down from one shift to the next aboard the Piper Alpha oil and gas rig. One of the pumps was shut down for maintenance and had its pressure relief valve removed. A work permit for this pump was neither clearly displayed, nor communicated during a shift change. When a blockage occurred in the other pump, the pump that was undergoing maintenance was put back online. This resulted in a gas leak and explosions that caused 167 fatalities.

Macondo, Gulf Coast, USA, 2010

See Appendix index entry C46

A cement barrier was improperly verified as effective. When the drilling mud was removed, this error resulted in hydrocarbons flowing past the blowout preventer (BOP). The release was eventually detected after nearly an hour had passed and the BOP was manually closed. Unfortunately, this was too late. The hydrocarbons reached the surface and found an ignition source. The failure of all these barriers resulted in 11 fatalities, 17 injuries, and approximately 4 million barrels of hydrocarbons released into the environment.

Montara, Timor Sea, 2009

See Appendix index entry S5

A small 'burp' of oil and gas was reported as having escaped from the well that kicked with such force that a column of oil, fluid and gas was expelled from the top of the well, through the hatch on the top deck, hitting the underside of the drilling rig. For a period of just over 10 weeks, oil and gas continued to flow unabated into the Timor Sea. Patches of sheen or weathered oil could have affected at various times an area as large as 90,000 square kilometers.

Bass Strait, Australia, 2012

See Appendix index entry S13

During an operation to attempt to free the drilling mechanism that had become stuck in the sea floor two crew members were struck by rotating equipment and suffered fatal injuries. Australia's National Offshore Petroleum Safety and Environmental Authority's (NOPSEMA) investigation concluded that on the morning of the incident, members of the drill crew were not given enough information to fully understand the amended plan of work and hence to carry out their roles safely. The Toolpusher and Senior Toolpusher failed to apply the company's management of change principles and failed to carry out a new risk assessment and toolbox talk after altering the original plan of works, and the Driller failed to ascertain whether a risk assessment had been carried out prior to implementing the new plan.

Charlotte also found several relevant onshore incidents where the process safety culture was broken:

Texas City, TX, USA, 2005

See Appendix index entry C11

The incident occurred during the startup of an isomerization unit when a raffinate splitter tower was overfilled; pressure relief devices opened, resulting in a flammable liquid geyser from a blowdown stack that was not equipped with a flare. The release of flammables led to an explosion and fire. All the fatalities occurred in or near office trailers located close to the blowdown drum. A shelter-in-place order was issued that required 43,000 people to remain indoors. Houses were damaged as far away as three-quarters of a mile from the refinery. Explosions and fires resulted in 15 fatalities. Another 180 were injured.

La Porte, TX, USA, 2014

See Appendix index entry C26

Approximately 24,000 pounds of highly toxic methyl mercaptan was released from an insecticide production unit that resulted in the deaths of three operators and a shift supervisor inside a manufacturing building. They died from a combination of asphyxiation and acute exposure (by inhalation) to methyl mercaptan.

CSB determined that the cause of the highly toxic methyl mercaptan release was the flawed engineering design and the lack of adequate safeguards. Contributing to the severity of the incident were numerous safety management system deficiencies, including deficiencies in formal process safety culture assessments, auditing and corrective actions, troubleshooting operations, management of change, safe work practices, shift communications, building ventilation design, toxic gas detection, and emergency response. Weaknesses in the safety management systems resulted from a culture at the facility that did not effectively support strong process safety performance.

Pittsburg County, OK, USA, 2018

See Appendix index entry C57

A blowout and rig fire resulted in the deaths of five workers who died from thermal burn injuries and smoke and soot inhalation. CSB determined the cause of the incident was the failure of both the primary barrier—hydrostatic pressure produced by drilling mud—and the secondary barrier—human detection of influx and activation of the blowout preventer—which were intended to be in place to prevent a blowout. The safety management system in place was ineffective. The operating company did not specify the barriers required during operations, or how to respond if a barrier was lost. Thus, employees were not effectively trained and ill-prepared to identify and respond to an incident.

11.3 Understand

Charlotte sat down with Lucas to discuss her findings. She summarized them and tried to relate them to her current work environment.

- *Piper Alpha; Bass Strait; LaPorte; Texas City*: All these incidents involved poor communication. For Piper Alpha and Texas City, there was a breakdown in shift communications. In Texas City, the plant did not have a shift turnover communication requirement for its operations staff. In the others, communication was fragmented, and management of change processes were not followed.

- *Macondo and Pittsburg County*: Several barriers were bypassed before the incident occurred. Human barriers are unreliable, particularly if there is lack of training or stress involved. Sleep deprivation could be one of the stress factors that results in slower reaction times and/or poor decision making.

- *Texas City and Montara*: Overall poor process safety culture led to incidents. The Texas City plant cut costs and did not upgrade equipment or infrastructure, leaving the refinery vulnerable to a major, potentially catastrophic incident. Furthermore, operators were likely fatigued from working 12-hour shifts for 29 or more consecutive days. For the incident in the Timor Sea, the report of the Montara Commission of Inquiry noted that the prevailing philosophy, revealed by the company's actions, appears to have been "Get the job done without delay," which resulted in risks not being recognized and poor decisions being made.

11.4 Drilldown

After her meeting with Lucas, Charlotte continued to think about all the information she had found. It was troubling to her to find so many instances where poor process safety culture resulted in the loss of so many lives. The Texas City incident really hit home for her. The 12-hour shifts for 29 or more consecutive days sounded all too familiar. Were there other shift arrangements that would be less stressful yet not impact production? Were there studies she could share with management that would help them to see the risk they were taking by demanding such a rigorous schedule? At the minimum, at least they would be informed about the decisions they were making. She identified two other questions her incident reviews suggested:

- One key driver of poor process safety culture is the lack of the sense of vulnerability. How could the company create this sense of vulnerability?
- Communication was a key factor in all the incidents she studied. What could the company do to ensure that written procedures were followed? How could they ensure that communication from shift to shift was done effectively?

11.5 Internalize

After thinking about the situation more deeply, Charlotte sat down with Lucas and Oliver to go over the findings and come up with a plan. Lucas invited Oliver to the meeting because he knew that an effective change in process safety culture needed to have complete leadership buy-in. If they could engage Oliver, they would have a stronger case when they presented it to Mason.

Charlotte started the meeting with a question, "Did you know that there have been safety incidents on other rigs under circumstances that are not very different from our own?" "Sure, but that's them," Oliver grumbled. "We're better than that. It won't happen to us." Lucas said, "I'm sure that's exactly what they thought too, until it happened.

Charlotte went on, "I reviewed the information on the Bass Strait incident where two men lost their lives. They didn't follow the procedures for management of change, likely because it would cause delays. There's no telling what would have happened that day, but the likelihood of this accident would have been reduced if they had just followed their own procedures." "You have a point," Oliver conceded. "We've been trying to avoid delays by finding workarounds, but we haven't gotten to the point where we don't follow our own procedures." Lucas drove the point home. "You don't want to wait until you get to that point," he said. "By then, it's too late. As the senior people on this rig, we have to remember that the decisions we make affect our crew."

Lucas said, "Charlotte has some other findings that I think may interest you." Charlotte said, "The human fatigue survey that we conducted with your crew provided some interesting results. It's clear that the twelve-hour shifts are taking a toll on the crew. People are just not getting enough sleep." Oliver said defensively, "Tell me something I don't know." Charlotte said, "On my survey, I asked if people would be willing to work a longer rotation, but with shorter work hours. Say either eight or ten hours instead of twelve. The response was overwhelmingly yes. It's a win-win. They work fewer hours, but they stay longer, so there's more continuity in personnel."

"That's an interesting compromise," Oliver said. "But I need to get a more detailed picture as to how it would affect my crew and potentially production." Lucas responded, "We'll get a couple of scenarios worked out in the next few weeks, and then we can get back together again." Oliver nodded. "Sounds like a plan."

Lucas said, "One last thing. We really need a united front to bring all of this to Mason. Can we count on you?" Oliver said, "One step at a time. Let's see what you come up with, and then we'll talk. You've given me a lot to think about."

11.6 Prepare

Over the next few weeks, Charlotte and Lucas worked on a more detailed plan to show Oliver and then Mason. Charlotte worked on developing several scenarios for shorter shift hours but longer rotation. Lucas worked on ways to create a sense of vulnerability on the rig.

At the follow-up meeting, Charlotte said, "I looked into various shift scenarios and recommend that we move forward with the ten-hour shift." She based her recommendation on API Recommended Practice 755 (API 2019):

- Work sets shall not exceed 9 consecutive day or night shifts.
- There shall be 36 hours off after a work set, or 48 hours after a work set containing 4 or more night shifts.
- Shifts are routinely scheduled for 10 hours and holdover periods should not exceed 2 hours and, where possible, occur at the end of the day shift.

Lucas said, "Charlotte, there's a bit of a problem here. Two ten-hour work shifts only add up to twenty hours. What happens to the other four hours in the day?" Charlotte flushed red with embarrassment and said, "I guess in my eagerness to come up with a solution, I didn't think about that. I only ruled out the eight-hour work shift because it increased our crew complement by fifty percent."

Lucas said, "Let's not panic here. I think I have a solution. When I served in the Royal Australian Navy, we broke the overnight watch into two shorter periods we called the dog watches." Lucas went on to explain, "From midnight to 0400 we could have a small skeleton crew with work limited to pumping oil and station-keeping. It might add to the crew complement, but maybe we can assign them tasks that others do, and we'll come out even. I think this could work. What do you think, Oliver?"

Oliver looked at it carefully and said, "It doesn't look like a big change, but I'd like to try it out first before fully committing to it." Lucas said, "That's all we're asking for. I think you'll find that with more rest, your crew will operate more safely and more efficiently." Oliver said, "I hope you're right. To see if it makes a difference, I'd like to see at how this new shift pattern compares with the previous one."

Charlotte chimed in, saying, "That's a perfect segue into the next part of my proposed plan. We don't regularly do fatigue risk management surveys, yet it's one of the key elements in API 755. I suggest that we collaboratively develop targets, like percent overtime and number of open shifts, so we can measure the impact of the change. Then, we conduct the survey on an annual basis, rather than the ad hoc way we've been doing it." Charlotte continued, "We should also look at other factors, such as healthcare costs and safety data." Oliver said, "Looks like you've come thoroughly prepared. I'm comfortable bringing this to Mason."

Now it was Lucas's turn to talk about trying to keep safety in the forefront of people's minds. Lucas said, "Creating a sense of vulnerability is a big challenge, but it's imperative in keeping all of us safe on this rig." He continued, "Everybody retains information differently. Some people like to read, while others like to watch a video or listen to a talk. I suggest we develop various ways to remind people of what can happen if they don't operate safely. But first, we need some shock value to wake everyone up. I think watching a movie on the Macondo catastrophe could do that. We'd need to have a follow-up discussion to really get the message across." Lucas then laid out the rest of his plan:

- Continue safety meetings at the start of each shift but try to make them more engaging by adding a reading of a short excerpt from a one-page safety newsletter like the CCPS *Beacon.*
- Post the newsletter around the rig to remind people to operate safely
- Watch investigation analysis videos from various safety agencies over lunch or dinner.
- Create and distribute stickers for hardhats aligned to this emphasis.
- Create a safety slogan or jingle contest.

Oliver said, "This seems all reasonable. Let's go and pay Mason a visit and see what he thinks."

When the group approached Mason about their plans, the first thing he asked was how much this was going to cost. Oliver said, "It all depends on how the shift change works out. It could be that we save in healthcare costs, and we'd save in transportation costs. We wouldn't have to fly people out to the rig as frequently." Lucas said, "There's a lesson to be learned from the Texas City incident. They cut costs and didn't invest in their plant. In the end, they had fifteen fatalities, nearly two hundred injured, and financial losses exceeding one-and-a-half billion Euros." After much discussion, Mason agreed to the plan.

11.7 Implement

Oliver scheduled several meetings with his crew to go over the new shift arrangement. He didn't want any confusion as to who was doing what. He made it clear that this was a pilot, and if everything went well, then it would become permanent. Crew members were pleased to see that the survey they took had some meaningful impact. Although they would be away from home longer, they would also be able to stay at home longer. Oliver said to the crews, "My door is always open. If you have any concerns about this new arrangement, don't hesitate to say something. The point of this pilot is to get you the rest that you need so that you can operate at your best."

As the safety officer, Lucas called for several meetings with the crew to roll out the safety communications plan. His first step was to have everyone watch the 2016 *Deepwater Horizon* movie. It was controversial, but it woke everyone up and brought safety into the conversations people were having on the rig. Oliver also instituted the reading of a safety briefing at safety meetings. It wasn't always the most effective, but it was one way to get people personally engaged. Posting the newsletter around the rig was a passive form of communication, but it took little time to do.

The last part of the plan was probably the most fun for the crew—creating a safety slogan or jingle. The valuable prize for the winner was additional time off. When it was time to judge the entries, there were so many to choose from. Some were very straightforward: "If you don't think it will happen to you, find the person who had it happen to them." Others were catchier: "If you mess up, 'fess up." Some were downright morbid: "Arms work best when attached to the body."

Lucas also received some musical suggestions. Although none of the crew could carry a tune, many of them remembered the new wave hit "Safety Dance," by Men Without Hats (Doroschuk, 1982). He also had a few people suggest a search on YouTube for safety songs. He had a good chuckle over a safety rap and decided he would suggest that Oliver occasionally play a YouTube video at the safety meetings instead of having someone read a newsletter out loud. Lucas was so pleased with the results that he decided to run the contest every quarter, but with a prize of a dinner for two, instead of time off.

11.8 Embed and Refresh

A year later, the pilot was deemed successful. There was a positive change in the attitude of the crew and no impact on operational efficiency. Lucas said to Charlotte, "I'm glad you took the initiative to try to make change. Sometimes, it takes a fresh pair of eyes to bring in a new perspective." "So...does that mean I'm getting promoted?" Charlotte joked. Lucas laughed and said, "Keep this up and you most certainly will be."

Charlotte smiled. "Now that a year has passed, it's time to do the fatigue assessment. It'll be easier this time since Oliver and Mason have bought into the value it brings," she said. Lucas said, "True, but don't rest on your laurels. We have to keep reminding people it's important to operate safely." Lucas then started to hum the new song he'd heard at the morning safety meeting—the winner of the latest safety jingle contest. "It's been stuck in my head all morning," he said. "Our crew is more musically talented than I thought!"

11.9 References

If so indicated when each incident described in this section was introduced, the incident has been included in the Index of Publicly Evaluated Incidents, presented in the Appendix.

Other references are listed below.

11.1 API (2019). Fatigue Risk Management Systems for Personnel in the Refining and Petrochemical Industries, 2nd Edition. API RP 755.

11.2 Dorschuk (1982). The Safety Dance [Song]. From the album "Rhythm of Youth" by Men Without Hats.

12
REAL MODEL SCENARIO: OVERFILLING

"Don't let your learning lead to knowledge. Let your learning lead to action."—Jim Rohn, Entrepreneur, Author, and Motivational speaker

> The individuals and company in this chapter are completely fictional.

Gouda Terminal is located along the Nieuwe Maas River in Rotterdam, Netherlands. The terminal is a main distribution hub that not only serves the Netherlands, but also Belgium and Germany. Its central location allows for petroleum and chemical products to be loaded from trucks, rail tank cars, and marine tankers. The terminal stores crude oil, gasoline, jet fuels, and diesel fuel from the Middle East, North Sea, and Russia. The facility consists of 25 steel aboveground, atmospheric storage tanks with a total capacity of over 400,000 m^3. The petroleum products are piped in through an automated flow control system.

Float-and-tape gauges are installed in all tanks to measure the actual level in the tanks. These mechanical gauges have been in use for decades and operate via a pulley system consisting of a large float inside the tank attached to a counterweight by a perforated tape. The liquid level readings are displayed in a unit located on the side of the tank. This unit is equipped with electronics that transmit level data to the company's inventory management system, allowing for remote monitoring of tank levels. The company has also installed two high-level alarms in each tank.

The operators and shift supervisors work in three shifts: 7 a.m. to 3 p.m., 3 p.m. to 11 p.m., and 11 p.m. to 7 a.m. At the beginning of each shift, the operators manually record the levels of each tank and report this to the planning and operations team. The members of this team are responsible for monitoring the materials and levels in the tanks. The Planning & Operations team is also trained to do manual calculations if there are any anomalies in the automated systems. The team manager is a stickler for making sure that her employees understand the meaning of numbers that come out of a computer and don't just take the numbers at face value. The Planning & Operations team also provides instructions to the operators on what needs to be done for the day. The relationship has been a smooth one, as there is mutual respect between the operators and the team.

Several years ago, a severe thunder and lightning storm disabled several of the tank gauges. This resulted in inaccurate tank-level readings. Fortunately, one of the operators who was taking the daily manual recordings of tank levels noticed the malfunctioning gauges and notified management before it became a greater problem. After an internal review, the company installed an automatic overfill prevention system. Despite the improvements in the safety systems and maintenance procedures, the terminal recently faced a spate of tank overflow issues after experiencing some severe weather.

12.1 Focus

Jan, the president of the company, walked into the room. "Anyone care to fill me in on what's happening with these tank overflow problems?" he said to Frederik, the shift supervisor, and Pamela, the operations manager.

Frederik spoke first. "My team of operators is following the maintenance procedures exactly as written, so I'm hoping that Pamela has some answers." Pamela looked at Jan and said, "My team has been double-checking our tank-capacity calculations and everything seems to check out. But I think with the extreme weather we've been experiencing over the last few months has compromised our level-monitoring equipment, giving us inaccurate readings."

Jan looked thoughtfully at Pamela. "Well, it is rainy season now, so there's not much we can do about Mother Nature," he said. "We've been able to weather the storms in the past, what makes these storms different?" Frederick piped up to say, "Unlike the past, these storms have been more intense. My operators have had to deal more frequently with downpours that last for hours and cause flooding." Jan responded, "Sounds like we need to do more research to get to the bottom of this problem. Get a team of your best and brightest, such as they are, together and come up with a proposal on how to resolve this issue. We don't want to wait for a big incident to occur before we take action."

After the meeting was over, Frederik and Pamela walked over to the coffee machine together, discussing next steps. They understood the urgency of resolving the overflow issues. Frederik said, "I'm going to ask Reed to join our group. If you'll recall, he was the operator who spotted the problematic tank gauges that led to the installation of the automatic overfill prevention system." Pamela nodded approvingly. "Reed would be perfect for this team. He's very smart and he knows our facility like the back of his hand. He's been with us for decades." Pamela thought for a moment, then said, "I'm also going to invite Alexandre to join us. He's incredibly detailed and could really help us out with data mining from past incidents." Frederik said, "Perfect. We have our team, now let's get to work."

12.2 Seek Learnings

Alexandre was eager to show Pamela that he was a rising star in the company. He agreed to do a comprehensive search for external incidents as well as a review of internal incidents and near-misses. Alexandre was already familiar with the Buncefield incident where a tank was overfilled, resulting in a vapor cloud that ignited, injuring more than 40 people. He had recently refreshed his memory about this incident by reading the article "Buncefield: Lessons learned on emergency preparedness," in the April 2017 issue of IChemE's *Loss Prevention Bulletin*. Being technologically adept, Alexandre quickly compiled a list of other tank overflow incidents from the index included in the CCPS publication *Driving Continuous Process Safety Improvement from Investigated Incidents*:

Kanagawa, Japan, 1996

See Appendix index entry J184

Crude oil spilled from piping at a crude oil storage terminal in Kanagawa prefecture. The crude oil leaked from an opening caused by corrosion of a small-diameter pipe from salt breeze and splashing seawater. External corrosion was overlooked because the piping was re-painted without completely removing the old paint. Beneath the old paint, corrosion continued and made an opening without drawing the attention of inspectors.

Buncefield, Hertfordshire, UK, 2005

See Appendix index entry S3

Safety systems designed to prevent tank overfilling failed and gasoline began to spill from the vents on the tank roof. A flammable vapor cloud accumulated over the tank farm and eventually ignited, resulting in an explosion that injured 43 people. The fire was not fully extinguished for several days. Many nearby businesses were severely impacted from the incident.

Bayamón, PR, USA, 2009

See Appendix index entry C15

An overfill of a 19,000 M^3 (5 million-gallon) atmospheric storage tank with gasoline caused a vapor cloud that ignited causing multiple tank explosions and fires. 17 of 48 tanks were burned and it took 3 days before the fire was under control. 3 people were injured.

Pamela had also mentioned her concern over climate change, so Alexandre did another search on process safety incidents caused by natural disasters. He discovered a CCPS *Beacon* on this topic that covered wildfires near Fort McMurray, Alberta, Canada; Hurricane Harvey along the Texas, USA, coast; and the earthquake and subsequent tsunami in Fukushima, Japan (Figure 12.1). Alexandre decided that of the three, the incident caused by Hurricane Harvey was the most relevant. He searched the CSB database and found an incident investigation in Crosby, TX.

CCPS

Process Safety Beacon

This Issue Sponsored by ioMosaic

Messages for Manufacturing Personnel
http://www.aiche.org/CCPS/Publications/Beacon/index.aspx

January 2018

Is Your Plant Prepared for a Natural Disaster?

Natural disasters can significantly impact or threaten process plants. In recent years, many events have been in the news because of their impacts:

Images 1 and 2 — Satellite and ground photos of wildfires near Fort McMurray, Alberta, Canada, taken in May 2016, show the location of nearby oil processing facilities and the extent of the fire.

Images 3 and 4 — Hurricane Harvey, at near maximum strength in August 2017, caused widespread flooding and damage along the Texas coast.

Image 5 — Trailers containing organic peroxides at a process plant exploded because refrigeration systems were out of commission due to loss of electric power following Hurricane Harvey.

Image 6 — The Fukushima Daiichi nuclear power plant in Japan lost power in March 2011 as a result of a major earthquake and subsequent tsunami. Insufficient cooling caused three nuclear reactor meltdowns, hydrogen-air explosions, and the release of radioactive material.

What can you do?

- Know your plant's emergency procedures for natural disasters and understand your role in preparing for, responding to, and recovering from the event. The kind of disaster that can impact your plant depends on the location of your plant.
- Check that your emergency procedures consider the potential for natural events with little or no warning, such as earthquakes and tornadoes.
- Understand how areas of the plant you are responsible for could be impacted by a natural event, especially where there are specific process hazards. Review disaster response plans and check that they are thorough and complete for your work area.
- If you identify gaps in the existing plans, take your concerns to the attention of management so the plans can be improved.
- Recognize that employees may not be able to report to work after a natural disaster, and that workers at the site may not be able to go home. Be sure your plans consider these possibilities, as well as the potential for limited staffing.
- Confirm that emergency response plans consider workers who remain at the site during and immediately after a natural event. Personnel will need food, shelter, and communication methods, and roads and other public infrastructure may be out of service.
- Develop a personal emergency plan for yourself and your family for the kinds of disasters that can occur where you live and work. You will not be able to work effectively if you are worried about your family.
- Read the November 2005 and June 2011 Beacons for more advice on natural disaster preparation.

Be ready for natural disasters!

Figure 12.1. An issue of the CCPS *Beacon* focused on natural disasters.

Crosby, TX, USA, 2017

See Appendix index entry C1

Extensive flooding caused by heavy rainfall from Hurricane Harvey exceeded the equipment design elevations and caused the plant to lose power to its critical organic peroxide refrigeration systems. Consequently, the company moved the organic peroxides to refrigerated trailers on higher grounds. Unfortunately, the rain persisted, eventually forcing all plant employees to evacuate from the facility. A 1.5-mile evacuation zone around the facility was established.

About a week later, the organic peroxide products decomposed causing the peroxides and the trailer to burn. Despite being within the evacuation zone, a highway adjacent to the plant remained open since it served as an important roadway for the hurricane recovery efforts. The fumes generated by the burning organic peroxides enveloped the highway. 21 people sought medical attention from exposure to fumes generated by the decomposing products. More than 200 residents were evacuated and were finally able to return to their homes after one week.

And then, while reading the news, he learned of another climate change-related incident in Russia.

Norlisk, Siberia, Russia, 2020

A tank of diesel fuel at a power station ruptured, releasing 20,000 tons of diesel spilled into the Ambarnaya River (Ilyushina 2020). Preliminary indications were that the permafrost underlying the tank thawed, causing the foundation of the tank to sink unevenly. The stresses on the tank led to its catastrophic rupture.

12.3 Understand

Alexandre met with his teammate Reed to go through his literature search findings. Alexandre felt confident that he had exceptional knowledge of theory, but he knew Reed had the hands-on experience that could expand his understanding of the external incidents.

- *Kanagawa, Japan:* Pipeline corrosion caused a leak in the piping. Although corrosion wasn't the problem at Gouda Terminals, Alexandre and Reed took the lessons learned as the need to be thorough about maintenance inspections at their facility.

- *Buncefield, Hertfordshire, UK, and Bayamón, PR, USA:* Faulty equipment led to these incidents. Faulty level readings and a malfunctioning safety system resulted in the release of gasoline. The gasoline was aerosolized, creating vapor clouds that found an ignition source. Of all the incidents Alexandre and Reed reviewed, they concluded these were the most relatable to Gouda Terminal, which is set up in a similar way.

- *Crosby, TX, USA:* Rain and storm surge from a hurricane flooded the facility, disabling equipment critical for maintaining process safety. The facility had identified its process hazards subject to flooding after a hurricane 12 years earlier. But Hurricane Harvey caused much greater flooding than seen before.

- *Norlisk, Siberia, Russia:* Unusually warm weather caused the soil (permafrost) to soften, leading to structural failure of the tank. Melting of permafrost had been increasing in recent years, but the tank foundation inspection schedule had not been adjusted accordingly.

Reed mentioned that they should also include the internal incident where the automated tank gauges failed. He said, "I've been around this place for a long time, so when the heavy rains coupled with the lightning strikes hit our area, I knew to look around immediately to see if there were any issues." He also casually mentioned that the mechanical nature of the float-and-tape gauges resulted in greater maintenance needs. "Moving parts wear out after a while and are subject to the wrath of Mother Nature," he sighed. "And as the climate changes, these heavy storms may become worse and more frequent." Alexandre absorbed this information and put it in the back of his mind. He knew that this information was likely relevant to resolving the tank overflow issues. He just didn't quite know how it fit in yet.

Alexandre said to Reed, "This is a lot of information to digest. I think we need to let it settle in over a couple of days, and then get back together for a deeper discussion." Reed agreed, but said, "Just keep in mind that I'm working the third shift this week, from eleven p.m. to seven a.m. Hope that's not a problem." Alexandre responded, "No issues here. Pamela and Jan are keen to hear what we have to say, so I'm game for a little late-night work."

12.4 Drilldown

Later that week, Reed and Alexandre sat down for a one-hour session to dig deeper into their findings.

- *Buncefield, UK, and Kanagawa, Japan:* Both incidents had visual inspection requirements. In the Kanagawa, Japan, incident, the leaking pipe was covered by a coat of paint. In the Buncefield, UK, incident, visual inspection was the primary way to determine whether the drain valves were open or closed. Was there something at their facility that they weren't seeing? Or perhaps, was there something that operators were seeing out in the field, but didn't deem important enough to report, that could be the cause of the problem?
- *Bayamón, PR, USA, and internal incident:* Faulty equipment is a major problem in tank farms. Ensuring proper maintenance and doing regular checks has been a way to safeguard against incidents. In their previous conversation, Reed had talked about the mechanical nature of the automated tank gauges. Were other kinds of gauges available that would be reliable and not have so many moving parts?
- *Crosby, TX, USA:* The plant was prepared for 100-year and 500-year floods, but Hurricane Harvey was more than that; some reports said it was a storm that was likely to occur every 25,000 years, while others said every 500,000 years. Regardless, climate change is happening and severe weather events are increasing. What could the facility do to protect itself from these unthinkable events? What limits should they set on their process hazard analyses when it comes to flooding? Could any of the products stored in the facility be severely impacted by flooding?
- *Norlisk, Siberia, Russia:* Although permafrost was certainly not relevant to Rotterdam, the tanks could indeed be undermined by flooding. If the frequency of flooding increased, many of the terminal's tanks could be affected.

12.5 Internalize

Reed and Alexandre worked several nights to develop a detailed presentation for their respective bosses, Frederik and Pamela. The day finally came for them to present their work. Alexandre started the meeting by saying, "I want to thank Pamela for giving me the opportunity to work on this project. Being a fairly young engineer, I learned a lot about process safety that is not taught at university."

He then did his deep dive into the external extreme weather incidents he had found. He focused on the facts. Rotterdam is prone to flooding, but has built dikes, pumping systems, and engineering feats such as the Maeslant

Barrier at the Hoek van Holland to effectively deal with flooding. However, sea level has risen 20 cm over the past century. Alexandre said, "Climate change appears to be a big concern for us. While we aren't going to be hit by a hurricane like Harvey, our rainstorms seem to be intensifying, which could logically lead to more severe flooding. I recommend that we review our inspection policy for tank foundations. We should also review our process hazard analyses to determine what our procedures should be in the event of an unprecedented flood. My question to you is, how do you define unprecedented?"

Reed spoke up and said, "You don't know what you don't know, until it happens. That's what I learned from the incident that occurred in Crosby, TX. Who would have ever predicted that much rain in such a short period of time?" Pamela and Frederik smiled, perhaps a little bit nervously, at the comment. Pamela said, "Good point. We have to start somewhere, and ultimately, it will be Jan's decision on how far we go to protect the site against flood." Frederik offered, "Let's review the PHAs before we decide to do anything. We should also consider the frequency of these reviews." Alexandre quickly tapped the keys of his computer, making note of these action items.

It was Reed's turn to talk about the current issue at hand, the tank overflow issues. "After much investigating," he said, "the minor tank overflows were caused by the severe weather impacting the float-and-tape gauges, just as I suspected. We're fortunate that we have a good crew with many years of experience who know what to look for after a storm, but those guys are eventually going to retire. You can try to capture their knowledge, but sometimes, what one of us thinks is common sense isn't the same for others."

Frederik responded, "We do have a great team, but you're right, the future will be tough if we don't find and train the right replacements. Do you have any suggestions on how to handle this?" "I'm glad you asked," Reed said. "My good buddy Alexandre and I have been doing some research on alternatives to the float-and-tape gauges. We figured, we can't solve the workforce issue, but we can help you with gauges that aren't as susceptible to breaking down in extreme weather."

Pamela said "You've hit the nail on the head. Our plant relies on accurate level measurement as a key risk-reduction measure. But as you've mentioned, with the increased storm intensity, the current level indicators are increasingly losing their reliability. We need to find a way to maintain the same level of risk reduction."

Reed and Alexandre showed Pamela and Frederik a summary of the various gauge options (Table 12.1). Alexandre said, "I was very fortunate to find an article titled, 'Tank Gauging Systems Used for Bulk Storage of Gasoline,' in the Hazards XXII Symposium Series No. 156."

Table 12.1 Summary of Tank Gauging Systems

Method	Pros	Cons	Applications
Dip Tape	Low cost	Accuracy highly dependent on the skills of the person doing measurement	Calibrating other types of gauges Manual level measurement
Float-and-tape	Accurate enough for custody transfer	Phasing out of use, parts increasingly hard to find	Has been used extensively in fuel storage tanks
Servo-operated float	High accuracy	System has mechanical parts, becomes less accurate as they wear Maintenance costs can be high	Favored in gasoline storage industry
Radar tank gauges	No mechanical moving parts, reducing maintenance costs High accuracy	Correct initial set up of the radar system is difficult Can be prone to "false echo" measurements if tank contains internal fixtures	Favored in gasoline storage industry

Reed spoke up to say, "With Alexandre's skills finding relevant public information, and my field knowledge accumulated from experience and talking with other operators, we narrowed the field to two types of gauges, servo-operated and radar." He continued, "I have a few inquiries out to some of my buddies, asking if their companies made any upgrades on their tank gauging equipment. Once I hear back from them, I'll let you know what their first-hand experience has been with the upgrades."

Frederik said, "I should check with my peers at other companies to see what their experience has been as well." Alexandre followed up, "We were really attracted by the lack of mechanical parts in the radar gauges, but since the initial costs are so high, I still need to crunch some numbers to see if we can justify the expense."

Pamela and Frederik smiled broadly. Their employees had done a stellar job analyzing the situation and making a sound proposal. "Looks like we have the start of a plan that we can lay out to Jan for feedback and, hopefully, approval. We just need to flesh out a few more details and we should be good to go. Let's plan on getting back together in another week to finalize the plan," said Pamela. Before closing the meeting, Alexandre read back the action items, saying, "Here's what I have, and please let me know if I missed anything":

- Review current PHA to see if we need any updates to our emergency response to flooding. Check on the frequency in which we review PHAs. (Pamela and Alexandre).
- Follow-up with colleagues about first-hand experience with tank gauge upgrades. (Frederik and Reed).
- Lifecycle cost analysis for tank gauge upgrade. (Alexandre).

Pamela said, "I think you've captured everything. Good meeting." They left the meeting excited about the opportunities to improve the safety and operations of the facility, but also concerned that the costs might be so prohibitive that they would be stuck with the status quo. Upgrading the gauges and planning for a flooding event that might never happen could be quite costly.

12.6 Prepare

Frederik, Pamela, Alexandre, and Reed filed into the meeting room. Reed started first. "Following up on my action item about talking to my friends, my buddy Pieter said his company recently upgraded to radar gauges." Reed went on, "Like Alexandre mentioned in our previous meeting, we were leaning toward radar because it has no mechanical parts, but I was concerned about the complexity of the set-up and how my fellow operators would accept this new type of equipment.

"Pieter said the initial set-up and calibration has its challenges," Reed continued, "but nothing that can't be overcome with the right personnel." Fredrik said, "That's what my peer Johan says as well. Like me, he is a shift supervisor at another facility. He told me that using radar technology requires operators with a certain skillset. Getting outside personnel to perform maintenance is costly and can add up quickly."

Pamela wasn't surprised to hear the feedback. For a new technology to be successful, employees must embrace the new technology and see it as something that will help with their jobs. Pamela just wasn't so sure Gouda Terminal was ready for this change. A significant number of operators were close to retirement age, and they were just beginning to start hiring the next generation of operators.

Pamela moved the discussion along, saying, "Let's hear from Alexandre about the lifecycle costs of radar gauges. "My lifecycle cost analysis had a lot of guesswork involved," Alexandre said. "In addition to the cost of the tank gauging system, I considered the cost of the physical installation, any tank modifications, system commissioning, possible loss of operations, operator training, and annual maintenance and calibration support. There was no clear-cut justification to upgrade or not upgrade. While the maintenance expenses are lower, the other expenses I listed can eat away at this savings. It really all hinges on what our staff is capable of handling."

Pamela said, "From what it sounds like, we need to assess the skills of the operators we currently have on staff and outline the skills we want for the operators we hire in the future. We should put together a long-term plan to switch over to radar gauges. But for the short-term, let's focus on how to best maintain the equipment we have today." Frederik, Alexandre, and Reed all nodded in agreement. Pamela said, "We should show Jan all of our findings on upgrading to radar gauge technology, and also come up with an improved process for how to check our tank levels after a severe storm."

Reed said, "I'd recommend that after a rainstorm, we do a double verification on the tank levels. We shouldn't just take the float-and-tape gauge reading at face value. We should also manually measure with dip tape. My fellow operators won't be too pleased, but they know how to do this." Frederik said, "I think it's a sound suggestion."

The group moved on to the final action items regarding reviewing tank foundation inspections and PHAs. Alexandre started the conversation by saying, "First, a little history lesson. The North Sea floods of 1953 and 2007 have resulted in our country shoring up its sea defenses to meet the safety level of a flood chance once every 10,000 years for the west and once every 4,000 years for less densely populated areas. The primary flood defenses are tested against this norm every five years."

"That's all very interesting, but how does that apply to us?" Frederik said. Alexandre responded, "I wanted to show the great lengths and expense our country is going through to protect us from flooding. Then, I wanted to follow it up with how we are aligned, with our processes set up to handle a flood chance once every thousand years. We're in good shape, but we have to make sure that when we revalidate our PHAs every five years, we check that our assumptions about flood protection remain valid."

They presented their findings to Jan the following week. Jan said, "Good work team. It's a comprehensive solution that addresses our problems today, while looking toward the future. I'm willing to invest in upgrades, as long as there are solid justifications for them."

12.7 Implement

Frederik developed a communications plan for the new procedure on tank level readings after a severe storm. The plan focused on how this added verification step was meant to make the plant safer, which meant keeping everyone safer. He took some notes during Alexandre's presentation on overflow incidents and planned to share a finding from the CSB, that "overfilling was cited as the most frequent cause of an accident during operation; among the 15 overfill incidents found, 87% led to a fire and explosion." He also planned to show a brief video on Buncefield and Bayamon to drive the point home.

With Reed's help, Frederik rolled out the new procedure to the three shifts, and although there was some grumbling, the operators all understood that this extra work was for the greater good of the organization, and perhaps more importantly, for their own safety. Showing the videos really helped get the message across. Sometimes, pictures, or in this case videos, are worth more than a thousand words.

Pamela, in the meantime, developed a list of skills that would be required of the new hires. She wanted to make sure that when they transitioned over to the radar gauges, there would be no issues around accepting the new technology.

12.8 Embed and Refresh

Three years had passed since Frederik, Pamela, Alexandre, and Reed came up with a plan to address the minor tank overflows. At Reed's retirement party, Jan said, "First, I want to thank Reed for all his years of service with us. He has been such an important part of our work family. We might not have resolved the overflow incidents a few years ago without his diligence and hard work. We will most certainly miss him." "Hear, hear!" the crowd cheered. Jan continued, "Second, I'm pleased to announce that we will embark on a new adventure, one that will modernize our terminals. We will put into motion a project to upgrade our tank measuring system with radar technology. Everybody will be critical to the success of this project, and that includes all the new folks, as well as the more seasoned ones, like myself."

Frederik, Pamela, Alexandre, and Reed looked at each and smiled. The plan they had put together three years ago was finally coming to fruition. Alexandre, who Pamela considered her protégé, was going to lead the project. "When I retire, I'll be leaving the department in Alexandre's capable hands," Pamela thought with satisfaction.

12.9 References

If so indicated when each incident described in this section was introduced, the incident has been included in the Index of Publicly Evaluated Incidents, presented in the Appendix.

Other references are listed below.

12.1 Ilyushina, M. (2020). Putin declares emergency over 20,000-ton diesel spill. CNN (4 June 2020).

13
REAL MODEL SCENARIO: INTERNALIZING A HIGH-PROFILE INCIDENT

"Being ignorant is not so much a shame, as being unwilling to learn."
—Benjamin Franklin, Scientist, Inventor, Philosopher, and Diplomat

Chana Oil Seed, Ltd., is a family-owned and family-run food products company in Thane, India, on the eastern edge of Mumbai. The original owner recently sold it to Medjool, Ltd., a multinational nutritional company headquartered in Dubai. Under the acquisition agreement, the Thane facility continues to operate as Chana Oil Seed India, Ltd.

The individuals and company in this chapter are completely fictional.

Chana's primary business involves extracting oil from native seeds and then purifying the oil to food grade. Using hexane as an extractant, Chana produces oil from canola, soybeans, sunflower seeds, safflower seeds, peanuts, and coconuts. Chana also operates a small-scale unit to partially hydrogenate a blended oil.

Over the years, Chana has had process safety incidents on a regular basis. Small hydrogen leaks (with their nearly invisible fires) happened regularly in the early days, requiring intense effort to detect them and tighten up flanges. By the time of the acquisition, hydrogen leaks had been reduced to about one to two per year.

13.1 Focus

With all the attention on hydrogen and product purity, the business has paid less attention to hexane. Samir, the new Emirati plant manager, expressed his concern about the number of hexane spills to Abishek, the new engineering leader. Several releases to the sewer had occurred in just the past year, with a similar number of hexane spills into containment dikes. Luckily, none of the spills had caught fire.

Samir and Abishek quickly worked out simple ways to address the obvious reasons for the spills. They assigned a longstanding Chana employee, Manoj, to make the engineering changes, such as interlocking tank inlet valves on high level, and adding an educator/lift system for emptying rainwater from dikes, removing the manual drain valves that were always being left open. They also invited Rakesh, an experienced process safety engineer, to interview for a new position focused on training and procedures, implementing Medjool's policies, and improving overall hazard identification and risk analysis efforts.

Rakesh's interview occurred in late 2017, not long after the CSB released its report on the refinery fire in Baton Rouge, LA, USA (See Appendix, Index Entry C31). During his plant tour with Manoj, Rakesh pointed out the wide diversity of equipment used to perform essentially the same role; a key factor in the Baton Rouge fire had been an older valve different from the current plant standard. Rakesh also noted that the isobutylene used at the Baton Rouge facility had similar physical and hazardous properties to the hexane used in the Chana plant.

At the conclusion of the interview, Samir offered Rakesh the position, effective the next morning. His first assignment would be to see what else he could learn from the Baton Rouge incident and recommend modifications or additions to Chana's hexane spill improvement plan.

13.2 Seek Learnings

Rakesh started by summarizing the key findings from Baton Rouge:

See Appendix index entry C31

Baton Rouge, LA, USA,2017

The plant used two types of gearboxes to operate its manual valves. Operators regularly experienced problems with both kinds of gearboxes, either gears becoming stripped or connection pins shearing off. When pressed for time, operators would sometimes remove the gearbox to open the valve with a wrench. With either type of valve, this could be done safely. However, the older design gearbox had two possible ways to remove the gearbox. One way was safe, but the other involved removing the bolts that also secured the valve bonnet. The safe way could have been verified with a simple Job Safety Analysis (JSA), but none was done.

When an operator removed the wrong bolts, the bonnet came off, causing a major isobutylene release and fire which resulted in four fatalities. This valve design met the API standard when the plant was built but the standard was updated in 1984 to require the gearbox to be separately mounted. The standard allowed for legacied valves to be considered compliant.

In his search for other relevant cases, Rakesh found many incidents in the public domain where a current standard needed to be upgraded. Chana's standards all needed to be updated. He was sure he would also find many incidents caused by not following standards. However, he could not find another incident caused by failure of a legacied process component.

He decided to focus on the other factors in the Baton Rouge incident and found case studies that represented each factor.

- *Generic procedure* (procedure didn't state clearly how to safely remove the gearbox):

  Port Neal, IA, USA, 1994

  See Appendix index entry S14

  A massive explosion resulted in 4 fatalities and injured 18 when an ammonium nitrate neutralization reactor exploded. Offsite releases of anhydrous ammonia continued for the next 6 days. The reactor's pH control system failed, but operation continued with manual sampling. Later, the reactor was idled due to a shortage of nitric acid feed. Because it was winter, operators attempted to keep the reactor warm, using a sparge of high-pressure steam instead of applying low pressure steam to the reactor jacket. The pH drifted low, destabilizing the ammonium nitrate. Then the temperature rose to the decomposition temperature and the ammonium nitrate detonated. Operating procedures did not match the equipment in the field, didn't specify clearly what to do during winter shut down, or how to control pH when the sensor was out of service.

- *Error traps* (routinely doing a task one way, but occasionally needing to do it a different way):

  Anonymous 2 (from chapter 10):

  Process Safety Beacon Mar. 2009

  A driver was directed by an operator to connect the hose for a shipment of sodium hydrosulfide to the wrong connection. This material was rarely delivered, and the operator thought the truck contained the more routinely delivered material ferrous sulfate. The two chemicals reacted, forming a toxic cloud of hydrogen sulfide that was fatal to the driver.

- *Safe work practices* (failing to perform a sufficiently thorough job safety analysis):

Aichi Prefecture, Japan, 1996

See Appendix index entry J34

Workers were repairing a walkway above a wastewater neutralization vessel. While using a grinder, sparks fell towards the vessel, leading to a vapor cloud explosion that injured 4 people. The wastewater tank had an upper layer of toluene, ethylbenzene, and other flammable materials. Sludge that had built up on the wall of the tank also contained organic peroxides that may have been decomposing, increasing the oxygen content in the atmosphere above the vessel. The person issuing the safe work permit assumed that the wastewater was just water and did not test the atmosphere. No precautions were required by the permit to prevent sparks from falling into the tank. Workers did not consider whether the atmosphere above the tank might be flammable.

- Normalization of deviance (accepting deviations from safe behavior on a routine basis):

Space Shuttle Incidents, 1986 and 2003

See Appendix index entries S9 and S10

Both incidents, described in more detail in Section 8.7, involved continuing and successive deviations beyond safe operating limits. In the 1986 incident, launches had been made at colder and colder outside temperatures below the minimum acceptable temperature. In the 2003 incident, the requirement that foam not separate from the external fuel tank had been ignored many times, despite continuing evidence of damage to the shuttle's wings from foam strikes.

- *Running to failure* (not doing preventive maintenance on the gearboxes):

Rakesh found many examples of running to failure. However, most were related to an inspection interval that was too long based on the corrosion rate of the chosen material of construction (CSB 2001; CSB 2011; CSB 2013; HSE 2006; HSE 2020 and Appendix, Index Entry J9), or deferring replacement too long CSB 2012). But then the following incident came up on a food industry blog that he followed:

> Healdsburg, CA, USA, 2020
>
> Mechanical failure of an access door at the bottom of a 367-M^3 (97,000-gallon) tank of high-quality Cabernet Sauvignon led to loss of all but the tank heel (KQED 2020). Wine flowed into ditches and containment ponds in the winery before overflowing to the Russian River. Pumper trucks were able to collect some of the spilled wine from the ponds and ditches, but at least half made it to the river. Environmental damage had not been assessed at the time, but the retail value of the lost inventory was likely USD 10–20 million.

Figure 13.1 Example wine tank access hatch (Source: Wikimedia Commons)

13.3 Understand

Rakesh reported to Samir and Abishek that he hadn't found any external incidents other than Baton Rouge caused by legacied designs. But he did share what he had learned from incidents with similar contributing factors:

- *Port Neal*: No operating procedures existed specifying how to safely conduct manual sampling and pH adjustment when the pH control system was out of service. Also, no operating procedure existed for keeping the reactor warm during a cold-weather hold.
- *Anonymous 2:* The operator assumed the shipment contents were the same as he always saw rather than verifying the shipment and connection point.
- *Aichi*: The wastewater supervisor did not know the process well enough to recognize that flammable vapors could be present. Also, whether he knew the process or not, sampling for flammable atmosphere before hot work and repeated sampling regularly during the work period were supposed to standard procedure.
- *Healdsburg:* A critical component protecting valuable inventory was operated to failure.
- *Shuttles:* Near-misses were either ignored or accepted as successes because there was no loss.

13.4 Drilldown

As an experienced process safety professional, Rakesh knew well the concept described in CCPS Vision 20/20 of "Disciplined Adherence to Standards (CCPS 2020), especially as it applied to equipment designed to an older version of a standard. He clearly understood that even if the equipment, process, or practice has been legacied, the company must still determine if that process, practice, or standard manages risk adequately. With that in mind, he felt confident to skip the drilldown on the Baton Rouge incident and proceeded to examine the additional supporting cases.

- *Port Neal:* Reading further, Rakesh discovered that not only was the automatic pH control out of service, the jacket coils that were supposed to keep the reactor warm during shutdowns without overheating had also failed. Essentially, the site operators had allowed failures to accumulate and were improvising to keep running; this was classic normalization of deviance.

- *Anonymous 2 and Aichi:* The incident investigations did not describe the training the operator had received or how the layout of the systems could have fooled the worker into doing the wrong thing. Both factors could have contributed to the incident. But in any case, not verifying the material being unloaded and not sampling for a flammable atmosphere are both failures of operational discipline.

- *Healdsburg:* Rakesh had previously worked in a brewery and was familiar with the type of hatch door used in the industry. Did the hatch seal go bad? he wondered. That would have caused a slower leak. Normally, the spring tension would provide enough resistance to keep the hatch shut, so perhaps the spring failed. Perhaps, he thought, the quick-release handle snagged on the belt or pocket of a passing worker, providing a much larger opening for the wine to drain out. Either way, a critical component that put the entire process at risk had not been maintained. At his former brewery, they had changed their procedures (from car-sealing the valve handle to locking it) to protect the handle from snagging on pallets of product being moved by forklift.

- *Shuttles:* In both cases Rakesh reviewed, NASA was under tremendous political and economic pressure to launch. This pressure drove leaders to minimize the safety concerns raised by members of the program team and continue to operate under unsafe conditions. Rakesh recognized the conflict between production and safety when he saw it.

13.5 Internalize

Rakesh discussed his findings with Samir and Abishek as they walked through the plant. Workers were scurrying in all directions, shouting instructions to each other while supervisors urged them to hurry up. They observed a group of three workers, looking toward the ceiling, arguing and gesturing wildly. As they approached, they noticed a puddle on the floor and smelled the hexane.

"We're trying to figure out where the leak is so we can clamp it," said one of the workers. "No problem, sir," added the second worker. "It'll take just a minute."

Samir called over Prasad, the plant superintendent. "There are going to be some changes in how we operate," he said. "Shut down now. I expect you to shut the process down any time it gets out of the safe operating window. That includes every process leak. Figure out the source of the problem and then come back to Abishek and Rakesh with your team's recommendation."

Prasad said, "Are you sure? A clamp can keep us running. If we take all that time, we will lose production." Samir replied, "If the pipe fails because it was totally corroded, we'll lose a whole lot more. This is how it's going to be, okay?"

Back in the office, Samir said to Rakesh, "You didn't say this in your report, but do you believe that the poor maintenance, operational discipline, and normalization of deviance that were factors in your case studies are present in our plant?"

Rakesh fidgeted nervously. "Yes, sir," he replied. "And..."

"Yes, please go on."

"We have few standards and policies, which we mostly don't follow, and no real safety culture."

Samir nodded. "But we will."

13.6 Prepare

Over the next weeks, Abishek met several times with Rakesh, Manoj, and Prasad to develop the action plan. In the first few meetings, they didn't make much progress, as they continued to cycle back to saying, "We have always done it this way," and "Management will never approve." But finally, they moved past these roadblocks and mapped out a plan.

First, they would need a general policy, owned by Samir and enforced by Prasad, clearly stating their process safety expectations. The policy would be the cornerstone of a more formal operating culture which would include better operating discipline. Supervisors would drive operating discipline and adherence to the culture rather than over-stressing production.

The four struggled with this last point, worrying that if they didn't push production, productivity would suffer. They expressed this concern to Samir when he stopped by to check on their progress.

Samir complimented them on the work they'd already done, and then addressed the question. "How much productivity do we waste looking for temporary fixes?" he asked. "We should more than make up the difference if we manage our operations in an orderly fashion."

13.7 Implement

Abishek's engineering team developed the new standard equipment designs, worked with procurement to designate approved vendors, and implemented the new designs during a 30-day turnaround. During the turnaround, Samir and Rakesh led workshops and discussions with the supervisors to coach them on the new policy and the desired culture, especially the increased focus on operational discipline.

Then they watched and provided support as the supervisors relayed the message to the workers. At first, the new focus on operational discipline felt uncomfortable, especially for the long-time employees. But Samir was insistent, walking through the plant regularly to encourage good behavior, correct risky behavior, and answer questions. After a few weeks, workers began to see the benefits of the new approach. They came to appreciate Samir's commitment, and they began to get excited.

Toward the end of the turnaround, the mechanics and operators received specialized training on the modified processes and equipment. As start-up approached, Chana Oil Seed felt like a completely new company.

13.8 Embed and Refresh

After the usual start-up headaches, Chana settled into the new routine of production and maintenance. The first challenge arose three months later. Rakesh was in the plant monitoring for hydrogen leaks when he noticed a group of workers looking toward the ceiling, arguing and gesturing wildly. He approached the group and noticed a puddle on the floor. It smelled like hexane.

"What should we do?" one worker asked.

"What do you think we should do?" he replied.

"They weren't serious about shutting down to fix it, were they?" another worker asked.

"One way or another, you're going to find out," Rakesh replied. "But if I were you, I'd shut down and fix the leak."

This scenario played out several times over the next few months. Typically, the problem was a flange that had been improperly torqued during construction or the installation of the wrong gasket. One time a supervisor told the workers to hang a bucket under a leak and keep running. When Prasad saw the bucket, he reprimanded the supervisor and observed him closely for the next few weeks.

Six months after restart, Rakesh deployed the new hazard analysis procedure. It included a short on-site version for use in issuing work permits, a more in-depth version for use in MOCs, and a detailed version for formal PHA. Several pilot sessions were held for each version, during which Rakesh sought input from the workers and supervisors.

One year following restart, the president of Medjool, Inc., paid a surprise visit to the Chana plant. He and Samir had arranged for a production stand-down during the visit to celebrate all of Chana's improvements. During a lunchtime feast, the president walked around with a videographer. He asked some employees to say a few words about their commitment to following procedures, analyzing hazards, or process safety in general. He asked others to talk about what working at Chana used to be like.

The videographers quickly stitched together the video clips into a "Before and After" video which was played later in the afternoon. The following week, a series of posters showing different workers and their safety quotes appeared around the plant. These were greeted with much celebration.

13.9 References

If so indicated when each incident described in this section was introduced, the incident has been included in the Index of Publicly Evaluated Incidents, presented in the Appendix.

Other references are listed below.

13.1 CCPS (2013). Moving from good to great: Guidelines for implementing Vision 20/20 tenets & themes. www.aiche.org/ ccps/moving-good-great-guidelines-implementing-vision-2020-tenets-themes (accessed June 2020).

13.2 CSB (2001). Tosco Avon Refinery Petroleum Naphtha Fire. CSB Report No. 99-014-1-CA.

13.3 CSB (2011). E.I. DuPont de Nemours & Co., Inc. Belle, West Virginia Methyl Chloride Release, January 22, 2010, Oleum Release, January 23, 2010, and Phosgene Release January 23, 2010. CSB Report No. 2010-6-I-WV.

13.4 CSB (2012). Chevron Refinery Fire. CSB Report No. 2012-03-I-CA.

13.5 CSB (2013). NDK Crystal Inc. Explosion with Offsite Fatality. CSB Report 2010-04-I-IL.

13.6 HSE (2006). Catastrophic failure of shell and tube production cooler. www.hse.gov.uk/offshore/sa_01_06.htm(accessed June 2020).

13.7 HSE (2020). Corrosion fatigue failure of tubes in water tube boilers. www.hse.gov.uk/comah/alerts/corrosion.htm (accessed June 2020).

13.8 KQED (2020). Nearly 100K gallons of wine spill from Healdsburg vineyard, reach Russian River. (23 January 2020).

14
REAL MODEL SCENARIO: POPULATION ENCROACHMENT

"Learn about the future by looking at the past.
—Tamil Proverb

> The individuals and company in this chapter are completely fictional.

The Qiezi Fertilizer Company manufactures custom blends of fertilizer. The facility was built in the early 1980s approximately 50 miles from Qingdao, China. The location is ideal, as nearby land is used to grow wheat, sorghum, and corn; plus, the railway adjacent to the property makes it expedient to transport chemicals to and from the facility. The facility stores various compounds including ammonium nitrate, potash, diammonium phosphate, and potassium magnesium sulfate.

Qiezi built the single-story, well-ventilated storage building with a steel frame and concrete floors. Although these materials were more costly than a wooden frame, the company was concerned about potential fire hazards. When the storage building was constructed, there were few residential and commercial buildings in the area. As time progressed, however, the demand for fertilizer skyrocketed. Sales soon went from thousands to millions of tons per year. The facility underwent a rapid expansion. Along with this growth came new employees who needed housing near the facility. The city of Qingdao was also rapidly growing, and this growth began to encroach on the facility site.

Unfortunately, urban sprawl has had some undesirable consequences for the company. Following the Tianjin warehouse explosion (Tremblay 2016), the Chinese government had announced that it will require select facilities to move to government-approved industrial sites, in an effort to improve public safety and environmental controls. The Qiezi Fertilizer Company is on the list because some of the chemicals stored onsite, notably ammonium nitrate, are in the nation's Catalog of Hazardous Chemicals and List of Hazardous Chemicals for Priority Management.

Moving a company site is complex, especially considering what chemicals are in storage. The company's supervisory foreman, who has environmental, health, and safety responsibilities, understands that moving is not as simple as loading everything onto trucks and moving out. Ammonium nitrate is stable under normal conditions but can undergo a highly explosive decomposition when heated in a confined space and mixed with combustible or incompatible materials. Fortunately, the company's management team has five years to figure out how to make the move. In the meantime, with the government's focus on public safety, it is imperative to design a communication plan that informs the key stakeholders of the company's activities and also how to respond if there is an incident.

14.1 Focus

Wai-Kee, the supervisory foreman, arrived five minutes early for the regularly scheduled Monday management meetings where the team reviewed the status of critical projects and brought up any concerns. As usual, Mei, the public relations and communications manager, was already in the room. Mei was always punctual about showing up to these meetings.

She noticed that Wai-Kee looked distracted. "You look worried Wai-Kee. What's on your mind?" She expected him to bring up the big move, so she was surprised when Wai-Kee said he was concerned about the new apartment buildings being built less than a mile from the facility. "With these homes so close, we really need to update our facility siting analysis and our emergency response plan," said Wai-Kee. Mei nodded in agreement. "You should bring that up during the meeting," she said. Just then, the rest of the management team arrived in the conference room.

Winston, the vice president of manufacturing, and Chen, the president of the company, sat down and called the meeting to order. After reviewing all the critical projects, Chen opened the floor to any topics of concern. Wai-Kee jumped in, saying, "I'm worried about the new housing being built so close to our facility. With such a strong emphasis on public safety, I think it would be in our best interest to update our facility siting analysis and emergency response plan." Mei agreed, emphasizing the importance of communication with the public. "We want to maintain harmony with our neighbors," she said.

Chen and Winston agreed with Wai-Kee and Mei. But Chen wanted more details. He said, "Come up with a plan including a timeline and what resources you think you will need. We'll put it on the agenda for next Monday's meeting."

Wai-Kee and Mei left the meeting and decided that they would meet at lunch every day to work on developing the plan. Wai-Kee was well versed in facility siting, having read CCPS's *Guidelines for Siting and Layout of Facilities* (CCPS 2018) and taken training courses to keep himself up to date. While he worked on the siting analysis, Mei developed communications plans for the public and for first responders. By the end of the week, they were ready to present their plan to Chen.

When Mei and Wai-Kee gave their presentation at the weekly management meeting, Chen and Winston were impressed by their thoroughness. They approved the plan and required that the two give regular updates at future management meetings.

14.2 Seek Learnings

With everyone on board, Wai-Kee enlisted the help of one of his new hires, Anna, who had recently graduated at the top of her class. Wai-Kee said, "Anna, I need you to review the public literature and see what we can learn from past incidents." Anna responded positively and made it a goal to review public databases from around the world for any ammonium nitrate explosions.

She found numerous incidents involving ammonium nitrate, many with severe consequences. Anna tabulated all the data she could readily find and summarized the more notable incidents in chronological order (Table 14.1).

Table 14.1 Ammonium Nitrate Incidents

Year	Place	Country	Site	Tons NH_4NO_3 exploded	Fatalities	Injuries
1921	Oppau	Germany	Plant	450	561	1,952
1924	Nixon, NJ	USA	Plant	—	18	100
1940	Miramas	France	Plant	240		
1942	Tessenderlo	Belgium	Plant	15	189	900
1947	Texas City, TX	USA	Ship	2,300	581	3,500
1947	Brest	France	Port	3,000	26	5,000
1972	Taroom, QLD	Australia	Truck	18.5	3	—
1988	Kansas City, MO	USA	Truck	29	23	—

Table 14.1 (Continued) Ammonium Nitrate Incidents

Year	Place	Country	Site	Tons NH_4NO_3 exploded	Fatalities	Injuries
1994	Port Neal, IA	USA	Plant	5,700	4	18
1998	Xingping, Shaanxi	China	Plant	27.6	22	56
2001	Toulouse	France	Plant	300	30	2,500
2007	Monclova, COA	Mexico	Truck	22	37	150
2013	West, TX	USA	Plant	—	15	260
2015	Tianjin	China	Port	800	173	798
2019	Camden, AR	USA	Truck	—	1	3
2020	Beirut	Lebanon	Depot	2,740	154[1]	6,000[1]

[1] As of August 7, 2020.

Anna summarized a few of these explosions:

Toulouse, France, 2001

An explosion occurred where about 200-300 tons of ammonia nitrate was stored (ARIA 2001). The explosion resulted in 31 fatalities and 2,442 people injured. The explosion leveled the facility and left behind a crater that was over 20-ft deep and 130-ft wide. The exact cause remains unknown.

Monclova, COA, Mexico, 2007

A truck carrying ammonium nitrate crashed into a pick-up truck and caught fire (Mattson 2007). The truck exploded, resulting in 37 fatalities and many more injured. The fatalities included three local reporters, four paramedics, three police officers and other bystanders who were watching the burning wreckage, unaware of the contents in the truck.

Tianjin, China, 2015

Nitrocellulose stored in a large warehouse spontaneously combusted after becoming overly hot and dry, resulting in a fire that triggered the detonation of about 800 tons of ammonium nitrate stored nearby (Tremblay 2016). The explosion resulted in 165 fatalities, including 110 emergency personnel and 55 nearby residents Another 8 were never found. Altogether, 798 people were injured. There was extensive damage to surrounding structures including apartment blocks and a railway station.

Camden, AR, USA, 2019

The brakes of a commercial truck hauling ammonium nitrate in Arkansas caught fire (Rddad 2019). The driver attempted to extinguish the fire, but with no success. The fire heated the content of the truck, which led to the explosion that resulted in the death of the driver.

In her search for information, she came across the May 2016 CCPS *Beacon* (Figure 14.1) that focused on ammonium nitrate incidents and mentioned the CSB report on the 2013 West, TX incident. She downloaded the Chinese version of the *Beacon* and the CSB video (CSB 2013). She decided to bring it with her when she presented her findings to Wai-Kee and Mei.

CCPS 化学工艺安全中心

工艺安全 警示灯 Beacon

提供给制造业人员的信息

本期由 AcuTech 公司资助

众多的重复事故！

2016年5月

2016年1月29日，美国化学品安全委员会（CSB）发布了一份事故报告和相关的动画视频。事故发生在2013年4月17日，德克萨斯的韦斯特（West）一家农用化学品贮存仓库发生了爆炸。事故共造成15人死亡，260多人受伤，整个仓库彻底被毁坏，周围社区也大面积受到影响。

仅仅几天之后，中国政府于2016年2月5日公布了2015年8月12日天津化学品仓库爆炸的事故调查报告，该事故共导致170余人死亡，700多人受伤，经济损失超过10亿美元。

这两起事故都发生在贮存有硝酸铵（AN）的仓库（硝酸铵是一种常见的化肥），而且仓库内同时还贮存有多种其它化学品。在这两起事故中，均出现了大火而把硝酸铵暴露在高温之下的情况。CSB在报告中明确地指出："在火灾情况下，硝酸铵表现出三种主要的危害：火势不可控制、分解产生有毒气体以及爆炸。"

你知道吗？

- CSB在德克萨斯州韦斯特的爆炸事故报告中，识别了自1916年以来的另外32起涉及硝酸铵的爆炸事故（包括2015年中国天津爆炸事故）。这些事故共导致过1500人死亡，数千人受伤。
- 在CSB列出的硝酸铵爆炸事故清单中，有1947年4月16日美国德克萨斯州德克萨斯城的格朗开普（Grandcamp）号轮船爆炸。该事故造成大约500人死亡，3000人受伤。这次事故被认为是美国历史上最致命的工业事故。

你能做什么？

- 你的工厂也许无需处理硝酸铵，也许你也接触不到其它类似的高危险性物料。然而，如果你要使用任何危险性物料，尤其如果你的工艺要在危险工况下运行，你就应该知道在你工厂里的这种物料和在这种工况下，过去曾经都发生过什么事故。
- 请工程师、管理人员以及有经验的员工介绍你工厂及其它类似工厂的曾经发生过的事故，要知道你的工厂都在做什么样的工作来预防类似的事故。
- 对于重复发生事故的其它案例，请参见2014年2月期和2016年2月期的《工艺安全警示灯》。
- 针对你工厂所涉及到的物料及工艺，上网搜索历史上所发生过的事故。

Figure 14.1 *Process Safety Beacon* in Chinese about Ammonium Nitrate

14.3 Understand

Anna met with Wai-Kee and Mei and showed them her research. She summarized her understanding of the incidents that she thought related to the situation at Qiezi:

- *Toulouse, France:* Although the cause was still unknown, the explosion was likely caused by lack of awareness of hazards. The ammonium nitrate likely exploded after mixing with incompatible and combustible materials.
- *West, TX, USA:* The facility was built with a wooden frame, and the bins storing the chemicals were combustible. The emergency response plan was lacking. The first responders were unaware of the dangerous content in the facility and did not respond appropriately. The nearby apartment buildings were destroyed, and more than 100 people were trapped in a nearby nursing home.
- *Tianjin, China:* The takeaway from this incident was that understanding the reactivity of other chemicals stored near ammonium nitrate is paramount. Ammonium nitrate is stable in normal conditions, but explosive when heated in a confined space.

Wai-Kee recommended that Anna also summarize the truck incidents. "When we re-locate our facility," he said, "we will probably move our chemicals via truck. We can get a head start on the safety of the move by putting this information together." The group dispersed after agreeing that they would meet again to dig deeper into these past incidents.

14.4 Drilldown

Anna looked for deeper findings and recommendations that might not be obvious from the investigated incident reports.

- *Toulouse and Tianjin:* A key step for Qiezi would be creating employee awareness of the hazards of ammonium nitrate and the conditions that could cause an accidental chemical reaction. Going one step further, was there a way to make ammonium nitrate less inherently explosive? Alternatively, was there an inherently safer chemical that could substitute for ammonium nitrate?

- *West:* Since the company couldn't prevent residential buildings from being built so close to the facility, what steps could they take to maintain public awareness? Should there be town hall meetings to remind people of safety? What additional barriers could be put in place to improve safety?

14.5 Internalize

Anna presented all her findings to her process safety colleagues Wai-Kee and Mei. She highlighted two things: education about hazard awareness to all onsite personnel as well as first responders, and effective communication with key stakeholders. The process safety team was well aware of the hazards of ammonium nitrate. "But what about the rest of the company?" she asked the group.

Wai-Kee admitted that with such a busy work schedule, it had been difficult to find time to develop effective new ways to remind people of the hazard. "We have safety moments at the start of every day," he said. "But if we keep focusing on only ammonium nitrate, the message will eventually fall on deaf ears."

As a group exercise, they all brainstormed ideas on how to keep things fresh and memorable for the onsite personnel. After the group exercise was complete, Wai-Kee asked, "How are we doing in terms of housekeeping? Is there anything we are mixing that is incompatible with ammonium nitrate? Are we keeping the place clean, so no dirt is tracked into the facility?" Wai-Kee was a stern taskmaster on good housekeeping procedures because ammonium nitrate is a strong oxidizer that can violently decompose when mixed with incompatible materials such as chlorates, mineral acids, or metal sulfides. Anna assured him that the housekeeping procedures were being followed. She also volunteered to check on the chemicals being blended to ensure that they were compatible with ammonium nitrate.

When it came time for the discussion on how to educate the first responders, Mei suggested, "Let's invite them to our facility once a month or once a quarter to review the chemical hazards." One of Anna's colleagues, Andrew, asked, "Is there a way for us to regularly provide them with a list of chemicals stored in our facility?" Wai-Kee said, "All good thoughts. We need to make sure they are aware of not only the ammonium nitrate, but also the other materials that we store. For example, they need to know that diammonium sulfate can release toxic ammonia gas when it decomposes at 154 °C (310 °F). First responders also need to know the locations of any water-reactive materials. Last thing we want is to accidentally make a situation worse." Anna collected all the ideas from the group and decided to review her notes later with Wai-Kee and Mei so that they could prioritize the changes that could be made with relative ease and have a high impact.

As the meeting progressed, Anna asked the group if they should cover any other topics. Wai-Kee said, "This discussion has been great for improving awareness and communication, but I think we should try to answer the question you posed in your in-depth analysis of the West, Texas incident. What additional barriers can be added to improve safety?" "Maybe we should review the facility siting analysis as a first step?" Andrew suggested. Wai-Kee thought this was a good idea. He knew that safety had been a big concern when the facility was originally built, but so much had changed since then. Wai-Kee nodded approvingly and closed the meeting, saying, "Let's go ahead with these ideas and see if there is room for improvement."

Seeing how much effort was going into making the current site even safer, Mei had an idea. "Do you think there is a way that we can convince the government that we are a safe and trusted neighbor, so that we do not have to move?" she asked. Wai-Kee said, "Let's bring that to Chen and Winston, and see what they think. Personally, I think we should at least try. We've been a good neighbor and have never had a serious incident since we were founded.

"I understand the desire to move us," he added, "but there are enormous safety issues in moving, not to mention the financial consequences."

14.6 Prepare

Wai-Kee, Mei, and Anna took a couple of weeks to follow up on the ideas presented at their meeting. One of the critical action items was to ask Chen and Winston for their feedback on the possibility of negotiating with the government to avoid a move. Wai-Kee was skeptical that they would receive a positive response, but much to his surprise, Chen said that he wanted to make every effort to convince the government that the move was unnecessary. "Moving should be our last option," he said. "If we can come up with ways to make the facility safer and show that we are proactive rather than reactive, we have a chance to change their minds." Given their marching orders, Wai-Kee, Mei and Anna quickly went to work on developing a plan.

Wai-Kee and Anna worked on improving the safety of the facility. Considering all the chemicals stored there, the explosivity potential of ammonium nitrate was the biggest hazard. Anna had confirmed that there were no incompatible materials and that good housekeeping procedures were being followed. But they decided to go one step further.

To convince the government that the company was committed to safety, they proposed to isolate the ammonium nitrate in its own blast-resistant building. This would minimize the possibility of inadvertently mixing ammonium nitrate with combustible or incompatible materials. The new building would be located in the remote corner of the site, as far as possible from the residential housing and equipped with a blow-out panel. To prevent new construction from coming close to the ammonium nitrate storage building, they proposed to buy the adjacent land; the company could use it to test new fertilizer blends on different crops. The cost of these changes would be significant, but much less than if the entire company had to move.

In the meantime, Mei and Andrew were working on the stakeholder outreach plans. They devised a plan that called for quarterly meetings with the first responders to review the hazards at the facility. They wanted to make sure the meetings were engaging and memorable. So, they proposed to hold a series of competitions for police officers and firefighters from the various districts to see which group could best handle a simulated emergency. For the police, the challenges would revolve around how to communicate with the public in the event of a situation and what actions to take: Calmly and effectively evacuate the residents to a designated shelter, or call for shelter-in-place? For the firefighters, the competitions would test whether they responded to fires or smoke with an understanding of the chemicals in the facility. As an added twist, they would also be challenged to minimize the environmental impact by using the least amount of water possible.

Mei also wrote into the plan a recommendation that the company sponsor an Agriculture Day, open to the public. The event would allow members of the public to get to know the company, plus safety could be discussed in an open environment. "The more informed the public is about what we do and what to do if an accident happens, the better," Mei said. Wai-Kee added that the employees involved in organizing Agriculture Day would be regularly reminded of safety when blending the fertilizers.

14.7 Implement

Chen was able to get a meeting with the party committee secretary, the governor of Shandong Province, and the mayor of Qingdao to make the case for not relocating the company. The government representatives were impressed by the plan, particularly the company's willingness to build a new building to house only the ammonium nitrate, but they needed more assurances that the public would be safe.

In the end, the government allowed the Qiezi Fertilizer Company to stay at its location provided it would limit the amount of ammonium nitrate in storage, upgrade its dust collection system, and allow for both scheduled and unscheduled audits. With the government's blessing, construction of the new building started and was completed in less than a year. Meanwhile, in the original building, the fans and blowers were upgraded to improve dust collection.

Mei and Andrew worked with the local firefighters and police to schedule quarterly meetings where a company representative would visit and talk about chemical hazards at the facility. They also collaborated with the government on establishing the inaugural Agriculture Day, where the highlight of the day was the first responders' competition. And they made it a priority to capture the highlights of the competition and share them through various media outlets, so that the message of safety would reach beyond those who attended the event. The short videos and memes would also serve as a reminder to residents long after the event was over.

14.8 Embed and Refresh

Three years later, Wai-Kee, Mei, and Andrew were standing in the crowd at the opening ceremony of the third annual Agriculture Day. Several government officials were giving opening remarks about the importance of the agricultural industry. As at the previous two Agriculture Days, many neighbors who lived near Qiezi had turned out with their children to feast on the delicious and plentiful food and learn about the agricultural industry.

New this year was a partnership with the local agricultural university. "What a brilliant idea," thought Wai-Kee. The university can help us educate the public about the benefits of the agricultural industry, and we can recruit students to come work for us. He took a moment for reflection. When Anna had first come up with this idea, she had pointed out another benefit. "The university gets something out of the relationship too," she had said. "They can attract new students. It's a win-win partnership."

As it turned out, this was Anna's final idea before she made the tough decision to leave the company for another job that gave her a significant increase in salary and a management title. Wai-Kee had been sad to see her leave, but he felt fortunate to have had the opportunity to serve as her mentor. He'd known that Anna would do well in any company, and sure enough, she was already applying the safety lessons she'd learned at the Qiezi Fertilizer Company at her new company.

While Wai-Kee was deep in these thoughts, Mei and Andrew were discussing the first responders' competition, which they called "Mission Not Impossible," a play on the American movies starring Tom Cruise. Mei said, "I wonder who's going to win this year's competition?" Andrew replied, "I'm not really rooting for any particular team. It's just fun to watch how these teams think so creatively about how to handle a situation. I think this competition has been a great success with all of the engagement." Mei smiled and thought, "I couldn't agree with you more. At the beginning, there were only four teams. Now, there are more than ten teams involved. It's only going to get bigger as we continue to hold this event."

14.9 References

If so indicated when each incident described in this section was introduced, the incident has been included in the Index of Publicly Evaluated Incidents, presented in the Appendix.

Other references are listed below.

14.1 ARIA (2001). Explosion of ammonium nitrate in a fertilizer plant. www.aria.developpement-durable.gouv.fr/fiche_detaillee/ 21329_en/ ?lang=en. (accessed June 2020).

14.2 CCPS (2018). *Guidelines for Siting and Layout of Facilities*. Hoboken, NJ: AIChE/Wiley.

14.3 CSB (2013). West Fertilizer Explosion and Fire [Video].

14.4 Mattson, S. (2007) 28 die as dynamite truck explodes in Mexico. *San Antonio Express-News* (10 September).

14.5 Rddad, Y. and Snyder, J., (2019). Driver killed after fertilizer truck explodes in South Arkansas; Area evacuated after blast that was heard miles away. *Arkansas Democrat Gazette* (27 March).

14.6 Tremblay, J.-F. (2016) Chinese investigators identify cause of Tianjin explosion. *C&EN* (8 February).

15
CONCLUSION

"The person who refuses to learn deserves extinction."
—Hillel the Elder, Rabbi, Sage, and Scholar

As we discussed in Chapter 1, and as documented in many CCPS publications, the leadership and technical discipline of process safety provides significant business benefits (CCPS 2019a). These benefits include increased revenue, lower costs, improved productivity, the removal of obstacles to growth, protecting the company image, and enhanced shareholder value—all while improving leadership overall. If these benefits did not exist, however, we would still have the moral obligation and societal imperative to eliminate process safety incidents.

Intellectually, we have known for a long time that any root cause gap in our PSMS, standards, culture, or policies can lead to a range of incidents. We know that by eliminating any given gap, we can prevent a wide range of incidents. We also know that properly investigating incidents and near-misses is one of the most impactful ways of finding these gaps.

Yet incidents based on the same root cause gaps continue to repeat—both inside individual companies and among companies and industry sectors. We continually fail to convert what we know intellectually into practice. This outcome can be attributed to five basic factors:

1. *Failure to perform complete incident investigations.* This results in failure to discover key findings and to implement recommendations that address all root causes.
2. *Failure to learn from others' incidents*. By doing so, the company lose the opportunity to improve without having to suffer an incident.

3. *Failure to address recommendations from incident investigations.* This is a sure way to quickly forget what was learned from the incident.
4. *Failure of leadership to monitor ongoing performance to ensure improvements are sustained.* People do what their leaders expect—or perhaps more accurately, people don't do what their leaders don't seem to care about.
5. *Failure to institutionalize the lessons learned.* Without sufficient reminders, people tend to revert to old behaviors, and the original gaps reoccur.

The first factor has been amply addressed in incident investigation guidelines provided by CCPS (CCPS 2019b) and others. This book focuses on the last four factors. The CCPS book *Process Safety Leadership from the Boardroom to the Frontline* (CCPS 2019a) can also be a useful resource.

Learning from your company's incidents and near-misses is obviously important. And learning from incidents that occurred at other companies and in other sectors can accelerate a company's progress toward eliminating incidents. Because process safety incidents occur infrequently in most facilities, the opportunities to learn from your own incidents may be limited. By seeking out incident reports across the industry and across industry sectors, you greatly increase your opportunities to learn. Most importantly, we must use what we learn to drive continuous process safety improvement.

Forgetting lessons learned not only halts progress, it causes us to lose ground as our sense of vulnerability wanes and complacency grows. Yet forgetting is inevitable unless we, as individuals and as companies, take intentional steps to keep lessons learned alive. These steps, simple as they are, will always be more cost effective than re-learning after a repeat incident.

The process of learning from external incidents is like learning from internal incidents, but with a few key differences, most notably two steps included in the Recalling Experiences and Applied Learning (REAL) Model: *Drilldown* and *Internalize*. These steps are important even in companies with a broad range of manufacturing operations, where there are opportunities to learn across the company. In any case, the learning model presented in this book should be used by board members, senior executives, safety professionals, and everyone with process safety responsibilities to learn continuously, to inform, and to reinforce our process safety improvement efforts.

The more than 440 externally investigated incidents indexed in this book (see Appendix) amply identify findings and recommendations that address the vast majority of PSMS, standard, culture, and policy gaps we continue to see across industry. CCPS recommends that companies explicitly include learning from externally investigated incidents in their process safety improvement efforts.

Several publicly investigated incidents are so iconic and impactful that they should become part of the basic knowledge of everyone across the industry. Even if a person cannot recite the details, if they are in the industry, they should be able to share the collective sense of vulnerability. They should then use that sense of vulnerability to motivate safe work and dedication to completing all tasks with professionalism. Chapter 8 summarizes incidents such as Bhopal, Flixborough, and Piper Alpha that need to be part of our collective consciousness. If any of the gaps that led to these incidents currently exist within our operations, we should make it a priority to eliminate them.

Senior corporate leaders own the PSMS; they are accountable to assure performance and drive improvement. As a person with process safety leadership responsibilities, you must be the one to help senior leaders carry out their responsibilities. It is up to you to ensure that internal and external incidents are considered and evaluated, following the REAL Model:

1. *Focus*. Identify where improvement is necessary.
2. *Seek learnings*. Find external (and internal) incidents with learnings that can help the improvement process.
3. *Understand*. Become thoroughly familiar with the incidents and all relevant findings.
4. *Drilldown*. Look for findings that the external or internal investigator did not describe.
5. *Internalize*. Translate the key findings into proposed changes to the company PSMS, standards, and policies.
6. *Prepare*. Develop a detailed plan to implement the changes.
7. *Implement*. Make the planned changes, communicate, and train.
8. *Embed and Refresh*. Manage the changes as implemented and refresh communication and training to make the changes part of the culture.

Taking proactive measures is paramount to improve process safety performance and is an integral part of Risk Based Process Safety (RBPS) (CCPS 2007). Although overcoming the obstacles laid out in Chapter 3 may be challenging, it can be accomplished with professionalism and leadership commitment. We should look to success stories like the Dow example described in Chapter 1 for inspiration. This example proves that it is possible to improve process safety performance and embed the changes in the culture while continuing to prosper. If you look around your company, you will probably find your own sites and units that have managed to do this. That is what we're all striving for.

> *"Leadership and learning are indispensable to each other."*
>
> —John F. Kennedy, 35th US President

Driving improvement from incidents—internal and external—and retaining lessons learned is possible, beneficial, and essential. Let's make a safer future by driving these improvements. Individually, as companies, and as the industry, it's up to us.

15.1 References

15.1 CCPS (2007). *Guidelines for Risk Based Process Safety*. Hoboken, NJ: AIChE/Wiley.

15.2 CCPS (2019a). *Process Safety Leadership from the Boardroom to the Frontline*. Hoboken, NJ: AIChE/Wiley.

15.3 CCPS (2019b) *Guidelines for Investigating Process Safety Incidents*, 3rd Edition. Hoboken, NJ: AIChE/Wiley.

Appendix: Index of Public ly Evaluated Incidents

A.1 Introduction

This book advocates the value of driving process safety improvement by learning from publicly evaluated incidents. The CCPS Learning from Investigated Incidents subcommittee recognized that it could be challenging to find incident reports that apply to a specific situation. To make this search easier, the members of the CCPS Learning from Investigated Incidents Subcommittee members evaluated approximately 440 publicly evaluated incidents. For each incident, one of the subcommittee members identified what they believed to be the primary and secondary findings in terms of gaps in Risk-Based Process Safety (RBPS) elements, culture core principles, and causal factors. To control the size of the index, committee members tried to limit the number of primary and secondary findings to three each. Therefore, when reading an incident report, you may find additional findings that you can apply to the learning process. Their classification of these incidents forms the basis of the index formally presented in Section A.3.

The publicly evaluated incidents included in this index come from the following sources:

- The US Chemical Safety and Hazard Investigation Board (CSB)
- The Dutch Safety Board (DSB)
- The Health Safety Executive (HSE) of the UK
- Agência Nacional do Petróleo, Gás Natural e Biocombustiveís of Brazil
- NPO Association for the Study of Failure (ASF) of Japan
- Selected stand-alone incident reports

The index includes only those incident reports published before December 31, 2019.

Readers should bear in mind that each of the 17 individuals who contributed to the index were influenced by their career experiences. For any given incident, another person might have identified a different set of primary and secondary findings. Readers should therefore use the index only as a tool to identify incidents with potentially useful findings and not look to it for statistical information.

While the printed version of the index only allows readers to search on single root causes, culture core principles or causal factors, an electronic spreadsheet version of the index with additional capabilities is available to download. The spreadsheet version provides capability to:

- search for incidents involving multiple root causes, culture core principles, and causal factors
- search for incidents by reporting source, industry sector, and equipment type.

Additionally, the spreadsheet contains any explanatory comments made the CCPS subcommittee member who indexed it.

Download the spreadsheet from the CCPS website at:

www.aiche.org/ccps/learning-incidents

On opening the spreadsheet, enter the password:

CCPSLearning

CCPS will update the electronic version periodically to add incident investigations reported after 1 January 2020 as well as investigations from other sources. Readers are invited to visit the index webpage periodically to check for updates.

CCPS also invites readers to help index future incidents. The indexing form may be downloaded from the same webpage. Once the form is complete, please submit to CCPS by email:CCPS@AIChE.org.

A.2 How to Use this Index

1. Use Section 6.1 to identify the RBPS elements, culture factors, and causal factor area at your company or unit where improvements are needed.

2. Look up these elements, culture factors, and causal factors in the index, Section A.3. Note the relevant index numbers. The listed primary findings

are those that contributed most to the incident; the secondary findings contributed less but still may have learning potential.

3. Find the titles of the reports in the listings in Section A.4. We do not provide a web address foreach report because web addresses change from time to time. However, searching the organization's website should take the reader quickly to the relevant report.

4. Read the reports and follow the remainder of the REAL Model as described in Sections 6.2–6.8.

A.3 Index of Publicly Evaluated Incidents

Each of the 441 incidents indexed by the CCPS Learning from Investigated Incidents Subcommittee has been assigned a code, consisting of a letter followed by a one- to three-digit number. The letter refers to the collection of incident reports:

A   Agência Nacional do Petróleo, Gás Natural e Biocombustiveís of Brazil.
C   The US Chemical Safety and Hazard Investigation Board (CSB).
D   The Dutch Safety Board (DSB).
HA   Alerts published by the Health Safety Executive (HSE) of the UK.
HB   Bulletins published by the Health Safety Executive (HSE) of the UK.
J   NPO Association for the Study of Failure (ASF) of Japan.
S   Selected stand-alone incident reports.

The number refers to a unique report found in that collection.

This index is organized in four sections:

- Section 1. Codes for reports with potential findings related to most RBPS elements.
- Section 2. Codes for reports related to most CCPS Culture Core Principles.
- Section 3. Codes for reports related to many causal factors.
- Section 4. A cross-reference to contents of the above sections from many elements, core principles, and causal factors that were not directly indexed.

Once codes that may be relevant to your effort have been identified, go to Section A.4 to find the report title and how to obtain it.

Section 1. RBPS Elements (In Element order)
Process Safety Culture
See specific Culture Core Principles as applicable
Compliance with Standards - Primary Findings
A5 C3, C5, C6, C18, C20, C21, C30, C37, C38, C39, C45, C47, C56, C66, C69, C77 D18, D19, D25, D33 HA3, HA6, HB4 J61, J82, J146, J164, J179, J194, J202, J203, J209, J210, J212, J215, J224, J236, J269 S9, S15
Compliance with Standards—Secondary Findings
A4 C10, C17, C25, C29, C33, C35, C43, C44, C68, C72, C75 D32 HB5 J18, J27, J106, J201, J213, J229, J230, J237, J244
Competency—Primary Findings
A5 C12, C19, C22, C26, C41, C42, C57 J23, J25, J26, J28, J37, J38, J39, J41, J98, J143, J144, J149, J150, J151, J152, J154, J155, J168, J170, J173, J176, J177, J180, J181, J182, J184, J185, J186, J189, J190, J192, J193, J194, J196, J197, J216, J217, J218, J220, J222, J228, J241, J251, J259, J266, J269, J270 S9, S10, S12
Competency—Secondary Findings
A6, A7 C8, C9, C10, C23, C24, C43, C44, C45, C52 D32 HB4 J17, J18, J20, J21, J22, J24, J42, J50, J51, J54, J82, J96, J100, J101, J116, J127, J130, J132, J133, J134, J145, J158, J159, J160, J161, J163, J178, J187, J188, J195, J206, J214, J234, J236, J250, J252, J255, J271 S4
Workforce Involvement—Primary Findings
C58 J196 S10

Section 1. RBPS Elements (Continued)
Workforce Involvement—Secondary Findings
C17, C44, C71 J165, J173, J184, J251 S9
Stakeholder Outreach—Primary Findings
A11 C53 D21 HB4
Stakeholder Outreach—Secondary Findings
C38 D9 S4, S11, S12
Process Knowledge Management
See specific Causal Factor(s) as applicable
Hazard Identification and Risk Assessment (HIRA)—Primary Findings
A1, A4, A6, A7, A10 C2, C8, C9, C10, C12, C13, C17, C19, C21, C22, C23, C24, C25, C26, C27, C31, C32, C35, C36, C37, C40, C41, C42, C44, C48, C49, C50, C54, C59, C60, C62, C63, C68, C69, C70, C74, C75, C76, C77 D9, D42 J1, J3, J8, J10, J15, J16, J17, J19, J20, J21, J22, J23, J24, J25, J26, J28, J31, J32, J34, J35, J37, J41, J42, J43, J47, J51, J52, J53, J54, J57, J59, J60, J69, J71, J74, J76, J79, J80, J81, J82, J83, J86, J90, J94, J95, J96, J99, J100, J101, J102, J103, J104, J106, J108, J109, J110, J111, J112, J113, J121, J124, J125, J126, J127, J128, J129, J130, J131, J132, J133, J134, J136, J137, J138, J143, J144, J145, J147, J161, J163, J164, J167, J169, J170, J187, J200, J206, J212, J227, J228, J229, J230, J231, J237, J242, J255, J257, J259, J261, J268 S4, S7, S8, S11, S13, S14, S16
Hazard Identification and Risk Assessment (HIRA)—Secondary Findings
A2 C1, C7, C22, C29, C33, C34, C39, C45, C55, C58, C61, C64 D7 J6, J9, J11, J14, J18, J27, J29, J30, J39, J43, J49, J55, J56, J61, J66, J68, J85, J87, J89, J93, J105, J117, J123, J141, J142, J148, J151, J155, J157, J158, J159, J160, J162, J168, J173, J175, J197, J198, J205, J210, J211, J214, J218, J232, J233, J234, J235, J238, J260

Section 1. RBPS Elements (Continued)
Operating Procedures—Primary Findings
A1 C12, C20, C24, C28, C37, C40, C59, C70, C76 D20, D42 HB6 J11, J16, J18, J20, J30, J38, J39, J43, J43, J45, J48, J49, J50, J58, J65, J70, J78, J81, J103, J108, J120, J123, J124, J125, J127, J134, J146, J147, J149, J176, J181, J187, J200, J206, J208, J216, J217, J234, J235, J252, J255, J256, J257, J261, J263, J271 S8, S11, S14
Operating Procedures—Secondary Findings
C2, C11, C26, C39, C57, C58 HB9 J15, J28, J31, J52, J53, J54, J56, J57, J69, J74, J87, J98, J107, J110, J130, J131, J133, J158, J171, J174, J177, J180, J182, J188, J197, J228, J237, J246, J254, J264, J268, J270
Safe Work Processes—Primary Findings
A2 C8, C9, C10, C20, C22, C27, C44, C51, C54, C55, C67, C68, C71, C77 D18, D20, D32, D42 HA6 J7, J18, J37, J39, J40, J41, J42, J43, J46, J49, J50, J59, J64, J67, J79, J82, J89, J109, J114, J122, J148, J162, J165, J167, J174, J177, J183, J185, J186, J188, J190, J195, J197, J198, J199, J201, J203, J209, J223, J229, J242, J247, J253, J259 S1
Safe Work Processes—Secondary Findings
A1, A4, A5, A7, A10 C16, C31, C32, C39, C42, C43 HB6 J15, J17, J19, J20, J23, J25, J26, J28, J34, J47, J51, J76, J97, J101, J111, J118, J128, J143, J144, J146, J176, J228, J234, J237, J252, J258, J261, J267 S4, S7, S15
Asset Integrity and Reliability—Primary Findings
A1, A4, A6, A7 C8, C11, C14, C15, C16, C17, C21, C23, C25, C28, C29, C35, C51, C52, C69 D19, D25, D33 HA3, HA6, HB3, HB4, HB5, HB7, HB9 J3, J4, J16, J17, J18, J22, J27, J29, J33, J40, J44, J56, J62, J73, J77, J80, J84, J91,

Section 1. RBPS Elements (Continued)
Asset Integrity and Reliability—Primary Findings (Continued)
J121, J122, J128, J130, J132, J144, J145, J146, J148, J153, J156, J160, J166, J168, J169, J170, J171, J172, J173, J175, J176, J179, J182, J184, J186, J193, J194, J195, J199, J202, J203, J205, J207, J210, J212, J214, J215, J218, J226, J230, J231, J232, J233, J238, J239, J241, J249, J250, J254, J256, J265 S10, S12, S13, S16, S17
Asset Integrity and Reliability—Secondary Findings
A2, A5, A11 C2, C9, C31, C32, C39, C40, C41, C47, C64, C73 HA1, HA2, HA5, HA8 J2, J57, J78, J88, J99, J104, J110, J112, J113, J125, J131, J134, J139, J167, J181, J183, J189, J196, J209, J213, J227, J235, J240, J255, J258, J262, J269 S9
Contractor Management—Primary Findings
C22, C22, C57, C71 J46, J71, J75, J189
Contractor Management—Secondary Findings
A1 C11, C46, C58, C70, C77 J47, J76, J172, J179, J190, J194, J202 S3
Training and Performance Assurance—Primary Findings
C10, C43, C45, C47, C60, C77 J37, J42, J46, J61, J70, J76, J83, J85, J97, J108, J120, J121, J129, J131, J133, J134, J143, J145, J149, J152, J174, J180, J181, J185, J187, J190, J195, J196, J197, J206, J216, J221, J244, J247, J248, J254, J256, J261, J270, J271 S7, S15
Training and Performance Assurance—Secondary Findings
A1 C11, C12, C19, C20, C22, C23, C24, C36, C38, C44, C61, C69, C70 D9, D20 J19, J20, J21, J25, J26, J28, J38, J47, J48, J49, J50, J51, J52, J55, J58, J63, J65, J122, J123, J125, J128, J132, J148, J154, J157, J162, J171, J177, J178, J186, J208, J209, J210, J211, J217, J236, J257 S1, S14

Section 1. RBPS Elements (Continued)
MOC—Primary Findings
A1, A2, A4, A6, A7, A10 C22, C22, C25, C31, C36, C47, C50, C51, C61, C63, C70, C72, C76 D9 HB5 J12, J24, J45, J55, J72, J86, J90, J94, J102, J106, J119, J132, J139, J167, J214, J240, J255, J260, J264 S4, S7, S16, S17
MOC—Secondary Findings
C3, C7, C8, C62 J10, J14, J16, J26, J91, J111, J154, J205, J238, J239, J246, J250 S13, S14
Operational Readiness/PSSR—Primary Findings
C7, C11, C70, C76 HA10 J26, J40, J69, J90, J111, J146, J171, J174, J177, J178, J179, J183, J184, J188, J189, J192, J213, J243, J246, J249, J263, J269
Operational Readiness/PSSR—Secondary Findings
D9 HA3, HA6, HB4 J23, J62, J103, J124, J147, J170, J185, J210, J212, J216, J217, J251 S5
Conduct of Operations and Operational Discipline—Primary Findings
A2, A5, A10 C3, C11, C12, C18, C26, C43, C50, C57, C58 D9 J2, J19, J28, J38, J49, J50, J51, J52, J53, J54, J55, J56, J57, J58, J61, J63, J67, J70, J72, J73, J114, J127, J130, J147, J151, J165, J171, J174, J178, J180, J182, J183, J188, J190, J192, J208, J209, J211, J217, J243, J247, J248, J259, J262, J270, J271 S3, S4, S5, S13, S14
Conduct of Operations and Operational Discipline—Secondary Findings
A6, A7 C13, C15, C20, C24, C27, C28, C60, C76 D7, D19 J21, J22, J24, J25, J32, J35, J40, J64, J65, J75, J76, J91, J108, J109, J116, J119, J128, J129, J131, J133, J162, J163, J170, J176, J181, J184, J185, J186, J212, J237, J253, J261 S1, S10, S12, S15

Section 1. RBPS Elements (Continued)
Emergency Management—Primary Findings
A7 C1, C16, C17, C24, C30, C40, C45, C48, C49, C61, C62, C74 J44, J107, J109, J122, J244 S1, S3, S8, S16
Emergency Management—Secondary Findings
A2, A10 C7, C23, C29, C34, C35, C36, C38, C69, C71, C77 D33 HB4, HB7 J24, J88, J149, J150, J164, J172, J227, J229, J233, J239, J245, J271 S4, S11, S12,
Incident Investigation—Primary Findings
C20, C34, C52 J74, J83, J86, J245, J264
Incident Investigation—Secondary Findings
C14, C22, C25, C37, C39, C50 D7 J16
Metrics
Not indexed. Readers seeking to improve metrics can make more rapid progress by following references A.1–A.3 (API 2016; CCPS 2009; CCPS 2018)
Audits
Not indexed. Readers seeking to improve auditing can make more rapid progress by following reference A.4 (CCPS 2011)
Management Review and Continuous Improvement—Primary Findings
C16, C27, C44, C57, C61, C70 D19 J164, J166, J169, J205, J260 S9, S15
Management Review and Continuous Improvement—Secondary Findings
A10 C8, C11, C26, C38 D25 HA10 J27, J39, J42, J43, J58, J79, J82, J143, J168, J206, J207, J229, J237 S16

Section 2: Culture Core Prin ciples (In Principle order)
Establish the Imperative for Process Safety—Primary Findings
A5, A7, A10 C15, C16, C17, C18, C22, C44, C57, C70 D7, D21, D25 J4, J5, J29, J30, J32, J35, J44, J58, J61, J68, J70, J85, J90, J91, J114, J116, J143, J144, J148, J157, J164, J165, J166, J167, J170, J173, J174, J175, J176, J177, J178, J179, J180, J181, J182, J183, J185, J186, J187, J188, J189, J190, J192, J193, J195, J196, J197, J199, J212, J237, J240, J262, J270, J271 S7, S14, S16, S17
Establish the Imperative for Process Safety—Secondary Findings
A2 C5, C13, C14, C19, C37, C46, C51, C53, C54, C69 D33 HA3 J1, J2, J3, J7, J79, J86, J130, J132, J147, J158, J160, J161, J163, J171, J213, J239 S1, S5
Leadership
See RBPS Elements Conduct of Operations and Management Review, and other Culture Core Principles
Mutual Trust—Primary Findings
D21
Mutual Trust—Secondary Findings
D25 J164
Open/Frank Communications—Primary Findings
C24, C35 D21 J59, J63, J189, J214, J216 S9
Open/Frank Communications—Secondary Findings
A11 C30, C70 D25 J90, J164, J270 S10, S14,
Sense of Vulnerability—Primary Findings
A2, A4, A5, A6, A7, A10, A11 C11, C12, C13, C19, C22, C23, C24, C44, C45, C68, C69, C70, C71 D18

Section 2: Culture Core Principles (Continued)
Sense of Vulnerability—Primary Findings (Continued)
J34, J47, J49, J67, J90, J91, J96, J98, J133, J141, J143, J144, J145, J146, J147, J148, J149, J152, J154, J164, J165, J167, J170, J171, J173, J174, J176, J177, J180, J181, J182, J183, J185, J187, J188, J190, J194, J197, J205, J206, J207, J212, J213, J214, J215, J216, J217, J218, J233, J240, J242, J262, J269 S5, S8, S10, S16
Sense of Vulnerability—Secondary Findings
C10, C15, C20, C25, C28, C31, C39, C51, C53, C54, C55, C60, C64, C74, C76, C77 HA3, HA8 J25, J26, J32, J35, J37, J38, J39, J41, J52, J53, J55, J56, J57, J59, J61, J65, J99, J132, J156, J163, J169, J178, J199, J236, J251, J252, J253, J259, J260 S2, S4, S14, S15, S17
Empowerment to Fulfill Process Safety Responsibilities—Primary Findings
A4, A10 J15, J168, J183, J184, J190, J192, J194, J196, J270, J271 S9, S15, S16
Empowerment to Fulfill Process Safety Responsibilities—Primary Findings
C70 J78, J82, J90, J147, J149, J166, J182 S10
Understand and Act on Hazards and Risks
See RBPS element HIRA
Defer to Expertise—Primary Findings
A5 J31, J33, J54, J164, J168, J169, J188, J196, J218, J269 S9, S10,
Defer to Expertise—Secondary Findings
C4, C20, C44 HA10, HB4 J38, J53, J100, J182, J197, J215
Combat Normalization of Deviance—Primary Findings
A2, A7 C10, C11, C12, C14, C24, C45, C69, C70, C71 HA6 J9, J45, J50, J146, J150, J166, J170, J171, J174, J178, J179, J184, J187, J212, J215, J217, J218, J243, J247, J248, J262, J263, J271 S10, S12, S14, S15

Section 2: Culture Core Principles (Continued)
Combat Normalization of Deviance—Secondary Findings
A4, A6 C20, C21, C26, C27, C44, C68 D19, D25 J25, J27, J28, J42, J52, J65, J114, J167, J168, J173, J194, J195, J233, J236, J237 S2, S4

Section 3: Selected Causal Factors
Consequence Analysis—Primary Findings
A2, A5 C23, C24, C56, C69 D21 HA10 J119, J143, J146, J149, J154, J156, J164, J165, J173, J174, J181, J182, J195 S3, S16
Consequence Analysis—Secondary Findings
A10 C17, C37, C39, C44, C48, C49, C70, C73 D42 J37, J80, J86, J98, J101, J129, J132, J147, J152, J153, J155, J171, J176, J180, J196, J233, J235, J239, J260 S4
Corrosion Under Insulation—Primary Findings
J33, J207, J265
Corrosion Under Insulation—Secondary Findings
C25 J262
Dust Explosion Hazards—Primary Findings
C18, C37, C39, C63, C70, C75 J75, J79, J83, J95, J102, J128
Dust Explosion Hazards—Secondary Findings
C20 J152
Facility Siting—Primary Findings
C11, C45, C72, C73, C74 D7 J119 S1, S12, S16, S17

Section 3: Selected Causal Factors (Continued)
Facility Siting—Secondary Findings
C1, C14, C24, C52, C53, C63 HA10 J109, J129, J150, J157, J233, J239 S4, S7,
Gap in a Standard—Primary Findings
C4, C9, C22, C29, C30, C42, C43, C44, C45, C56, C64, C65 D7 J84, J91, J117, J232 S17
Gap in a Standard—Secondary Findings
C6, C34, C38, C39, C54, C70, C72, C74 D19, D42 HB3 J6, J40, J81, J206, J207, J218, J227, J237 S15,
Human Factors—Primary Findings
A7, A11 C34, C46, C60, C71 J1, J11, J13, J14, J50, J52, J55, J63, J65, J66, J75, J76, J81, J105, J107, J110, J132, J134, J140, J157, J158, J165, J166, J171, J174, J176, J177, J178, J179, J180, J181, J182, J185, J187, J188, J189, J190, J196, J197, J200, J204, J213, J217, J230, J234, J251, J252, J253, J254, J271 S2, S15
Human Factors—Secondary Findings
A2, A4 C10, C12, C13, C27, C38, C49, C51, C68, C76 D9, D32 HA3 J23, J28, J30, J35, J40, J41, J43, J45, J48, J61, J97, J108, J115, J128, J130, J131, J145, J147, J148, J149, J167, J192, J194, J221, J233, J236, J237, J256, J257, J270 S5
Reactivity Hazards—Primary Findings
C1, C3, C7, C12, C24, C32, C36, C49, C61, C62, C68, C73 J1, J5, J6, J7, J21, J24, J25, J30, J31, J32, J35, J38, J42, J43, J47, J51, J56, J66, J68, J69, J71, J72, J74, J77, J78, J83, J90, J92, J97, J98, J99, J100, J101, J104, J112, J113, J116, J118, J120, J121, J123, J124, J125, J126, J129, J130, J131, J132, J133, J135, J142, J143, J144, J145, J146, J147, J148, J149, J150, J151,

Section 3: Selected Causal Factors (Continued)
Reactivity Hazards—Primary Findings (Continued)
J155, J157, J158, J159, J160, J161, J162, J163, J169, J171, J187, J193, J197, J264, J267 S4, S11, S12, S14
Reactivity Hazards—Secondary Findings
C30, C48, C50, C63 J8, J10, J12, J39, J44, J45, J54, J55, J65, J81, J86, J103, J105, J108, J114, J117, J127, J137, J153, J156, J252, J262, J271 S7
Relief System Design—Primary Findings
A2, A5, A6, A10 C11, C59, C73 J48, J77, J153, J175, J192, J227, J258 S4, S16,
Relief System Design—Secondary Findings
C21, C37, C62 J10, J22, J44, J98, J116, J136, J142, J251
Safe Design/Error in Design—Primary Findings
A2, A4, A5, A6, A7, A10 C2, C7, C13, C14, C26, C34, C39, C45, C46, C49, C56, C69, C70, C76 D20, D32 HA1, HA2, HA5, HA8, HB4, HB5 J19, J20, J21, J23, J29, J48, J57, J60, J62, J68, J69, J78, J80, J83, J87, J88, J90, J91, J93, J97, J103, J104, J105, J106, J107, J110, J111, J115, J119, J123, J124, J125, J126, J131, J132, J133, J134, J145, J151, J153, J155, J156, J158, J159, J160, J161, J163, J167, J169, J173, J176, J179, J182, J202, J207, J210, J212, J214, J215, J216, J218, J219, J226, J227, J228, J232, J235, J236, J237, J238, J239, J241, J244, J245, J250, J257, J258, J260, J265, J268, J269, J270 S3, S8, S9, S10, S16,
Safe Design/Error in Design—Secondary Findings
C3, C11, C15, C16, C20, C23, C24, C29, C32, C33, C36, C37, C38, C40, C41, C42, C44, C47, C48, C50, C52, C75 HA10, HB6, HB7, HB9 J9, J15, J37, J41, J42, J43, J44, J46, J49, J58, J61, J63, J66, J75, J77, J81, J84, J85, J86, J89, J99, J100, J101, J112, J116, J117, J118, J122, J139, J144, J149, J157, J164, J183, J206, J252, J261 S11, S13

Section 3: Selected Causal Factors (Continued)
Safety Instrumented System (SIS) and Independent Protection Layers (IPL)—Primary Findings
C15, C31, C32, C70 HB6, HB9 J53, J216 S5
SIS and IPL—Secondary Findings
C21 J29, J31, J32, J43, J77, J78, J90, J97, J130, J175, J234
Transportation, Traffic, and Collisions—Primary Findings
C33, C38 HA10 J172 S15, S17
Transportation, Traffic, and Collisions—Secondary Findings
J7, J79, J87

Section 4: Additional Index Terms	
Index Term	See:
Confusion	Causal Factor *Human Factors.* Also see RBPS Elements *Operating Procedures, Safe Work Practices* and *Conduct of Operations.*
Containment of Spills	RBPS Element *Compliance with Standards, Conduct of Operations,* or *Emergency Planning,* as appropriate
Continuous Monitoring	RBPS Elements *Safe Work Practices, Operating Procedures, HIRA,* as appropriate
Control of Explosive Atmospheres	RBPS Element *HIRA*
Control of Ignition Sources	RBPS Element *Safe Work Practices*
Corrosion	*Asset Integrity*
Dead Legs	RBPS Elements *Management of Change* or *HIRA,* as appropriate
Decommissioning and Demolition	RBPS Element *HIRA*
Failure to Learn	RBPS Element *Management Review* and *Continuous Improvement*

Section 4: Additional Index Terms (Continued)	
Index Term	See:
Fatigue	Causal Factor *Human Factors*
Flame Arrestors	RBPS Element *Asset Integrity*
Freeze Protection	RBPS Element *HIRA*
Hazard Communication	RBPS Elements *Training, Workforce Involvement,* or *Process Knowledge Management,* as appropriate
Hiring	RBPS Element *Management of Change* (includes organizational change)
Hot Work	RBPS Element *Safe Work Practices*
Inherently Safer Design	Causal Factor *Safe Design,* and RBPS Elements *Compliance with Standards, Process Knowledge Management,* or *HIRA,* as appropriate
Intoxication, Alcohol, and Drugs	RBPS Element *Training* and *Performance Assurance*
Isolation	RBPS Element *Safe Work Practices*
Job Change	RBPS Element *Management of Change* (includes organizational change)
Job Hazard Analysis, Job Safety Analysis	RBPS Element *Safe Work Practices*
Line and Equipment opening	RBPS Element *Safe Work Practices*
Lockout/Tagout	RBPS Element *Safe Work Practices*
Maintenance Job Planning	RBPS Element *Asset Integrity*
Management	RBPS Elements *Management Review, Continuous Improvement,* and *Conduct of Operations,* and the *Culture Core Principles,* as appropriate
Non-routine Work	RBPS Elements *Safe Work Practices or Asset Integrity,* as appropriate
Organizational Change	RBPS Element *Management of Change*
Permits to Work	RBPS Element *Safe Work Practices*
Positive Material Identification	RBPS Element *Asset Integrity*
Process Control	RBPS Elements *Process Knowledge Management* or *Management of Change,* as appropriate

Section 4: Additional Index Terms (Continued)	
Index Term	See:
Production Pressure	RBPS Element *Culture Core Principles,* especially *Establish the Imperative for Process Safety, Empowerment to Fulfill Process Safety Responsibilities,* and *Normalization of Deviance*
Safety Data Sheets	RBPS Element *Process Knowledge Management*
Shiftwork	Causal Factor *Human Factors*
Staffing Insufficient	Causal Factor *Human Factors*
Stress	Causal Factor *Human Factors*
Structural Integrity	RBPS Elements *Compliance with Standards* and *Asset Integrity*
Supervision	RBPS Element *Conduct of Operations* and the *Culture Core Principles,* as appropriate
Tightening: Over-tightening and Under-tightening	RBPS Element *Asset Integrity*
Weather, Extreme	RBPS Elements *HIRA* or *Emergency Management,* as appropriate
Work Permits	RBPS Element *Safe Work Practices*

A.4 Report References

Once you have found the codes for incident reports that meet your search criteria, use this section to find the website address and report title of the organization that makes the reports available. This section is presented first by organization name (the letter part of the incident report code), and then by number.

In general, we have provided a web address only for the organization's main web page rather than providing a direct link to each case study. This is because direct web addresses to reports change whenever the organizations update their websites. We have provided web addresses for standalone special inquiry reports since there is no common address for them, but these addresses too may change. In most cases, an Internet search on any of the report titles in this index will take you directly to the report.

A. Agência Nacional do Petróleo, Gás Natural e Biocombustíveis (ANP) of Brazil Reports (Portuguese)

(See www.anp.gov.br/exploracao-e-producao-de-oleo-e-gas/seguranca-operacional-e-meio-ambiente/comunicacao-e-investigacao-de-incidentes/relatorios-de-investigacao-de-incidentes)

Code	Investigation
A1	NORBE VIII Drilling Offshore Platform
A2	Major Fire at P-20 Offshore Oil and Gas Platform
A3	West Eminence Drilling Platform S-69
A4	Large Fire at FPSO P-48
A5	Underground Blowout in the Frade Oilfield—Area 1
A6	Seabed Oil Exudation in the Frade Oilfield—Area 2
A7	Explosion at FPSO Cidade de São Mateus
A8	Incident at Alpha Star Drilling Platform SS 83
A9	P-34 Oil Platform list
A10	Sinking of P-36 Oil Platform
A11	LOPC on ORBIG Oil Pipeline
A12	Operator Fall into Oil Tank

US Chemical Safety and Hazard Investigation Board (CSB) Incident Reports (English)

(see: www.CSB.gov)

Code	Investigation
C1	Arkema Inc Chemical Plant Fire
C2	Acetylene Service Company Gas Explosion
C3	AirGas Facility Fatal Explosion
C4	AL Solutions Fatal Dust Explosion
C5	Allied Terminals Fertilizer Tank Collapse
C6	Barton Solvents Explosions and Fire
C7	Bayer CropScience Pesticide Waste Tank Explosion
C8	Bethlehem Steel Corporation Gas Condensate Fire
C9	Bethune Point Wastewater Plant Explosion
C10	BLSR Operating Ltd Vapor Cloud Fire
C11	BP America Refinery Explosion
C12	BP Amoco Thermal Decomposition Incident
C13	CAI / Arnel Chemical Plant Explosion

US Chemical Safety and Hazard Investigation Board (CSB) Incident Reports (Continued)

(see: www.CSB.gov)

Code	Investigation
C14	Carbide Industries Fire and Explosion
C15	Caribbean Petroleum Refining Tank Explosion and Fire
C16	Chevron Refinery Fire
C17	CITGO Refinery Hydrofluoric Acid Release and Fire
C18	Combustible Dust Hazard Investigation
C19	ConAgra Natural Gas Explosion and Ammonia Release
C20	CTA Acoustics Dust Explosion and Fire
C21	D.D. Williamson & Co. Catastrophic Vessel Failure
C22	Donaldson Enterprises, Inc. Fatal Fireworks Disassembly Explosion and Fire
C23	DPC Enterprises Festus Chlorine Release
C24	DPC Enterprises Glendale Chlorine Release
C25	DuPont Corporation Toxic Chemical Releases
C26	DuPont LaPorte Facility Toxic Chemical Release
C27	E. I. DuPont De Nemours Co. Fatal Hotwork Explosion
C28	Emergency Shutdown Systems for Chlorine Transfer
C29	Enterprise Pascagoula Gas Plant Explosion and Fire
C30	EQ Hazardous Waste Plant Explosions and Fire
C31	ExxonMobil Refinery Explosion
C32	First Chemical Corp. Reactive Chemical Explosion
C33	Formosa Plastics Propylene Explosion
C34	Formosa Plastics Vinyl Chloride Explosion
C35	Freedom Industries Chemical Release
C36	Georgia-Pacific Corp. Hydrogen Sulfide Poisoning
C37	Hayes Lemmerz Dust Explosions and Fire
C38	Herrig Brothers Farm Propane Tank Explosion
C39	Hoeganaes Corporation Fatal Flash Fires
C40	Honeywell Chemical Incidents
C41	Imperial Sugar Company Dust Explosion and Fire
C42	Improving Reactive Hazard Management
C43	Kaltech Industries Waste Mixing Explosion
C44	Kleen Energy Natural Gas Explosion
C45	Little General Store Propane Explosion
C46	Macondo Blowout and Explosion

US Chemical Safety and Hazard Investigation Board (CSB) Incident Reports (Continued)

(see: www.CSB.gov)

Code	Investigation
C47	Marcus Oil and Chemical Tank Explosion
C48	MFG Chemical Inc. Toxic Gas Release
C49	MGPI Processing, Inc. Toxic Chemical Release
C50	Morton International Inc. Runaway Chemical Reaction
C51	Motiva Enterprises Sulfuric Acid Tank Explosion
C52	NDK Crystal Inc. Explosion with Offsite Fatality
C53	Oil Site Safety
C54	Packaging Corporation of America Hot Work Explosion
C55	Partridge Raleigh Oilfield Explosion and Fire
C56	Praxair Flammable Gas Cylinder Fire
C57	Pryor Trust Fatal Gas Well Blowout and Fire
C58	Sierra Chemical Co. High Explosives Accident
C59	Sonat Exploration Co. Catastrophic Vessel Overpressurization
C60	Sterigenics Ethylene Oxide Explosion
C61	Synthron Chemical Explosion
C62	T2 Laboratories Inc. Reactive Chemical Explosion
C63	Technic Inc. Ventilation System Explosion
C64	Tesoro Refinery Fatal Explosion and Fire
C65	Texas Tech University Chemistry Lab Explosion
C66	Third Coast Industries Petroleum Products Facility Fire
C67	Tosco Avon Refinery Petroleum Naphtha Fire
C68	Union Carbide Corp. Nitrogen Asphyxiation Incident
C69	Universal Form Clamp Co. Explosion and Fire
C70	US Ink Fire
C71	Valero Refinery Asphyxiation Incident
C72	Valero Refinery Propane Fire
C73	Veolia Environmental Services Flammable Vapor Explosion and Fire
C74	West Fertilizer Explosion and Fire
C75	West Pharmaceutical Services Dust Explosion and Fire
C76	Williams Olefins Plant Explosion and Fire
C77	Xcel Energy Company Hydroelectric Tunnel Fire

Dutch Safety Board (DSB) Incidents (In Dutch, English, or Both)

(See: www.onderzoeksraad.nl/en/)

Code	Investigation
D1	Emission 1.3-Butadiene, Vopak, 1 February 2019
D2	Emission Ammonia OCI Geleen, 31 May 2018
D3	Power Outage, Shell Pernis, 29 July 2017
D4	Vinylchloride Emissie, Shin-Etsu Locatie Pernis, 17 Mei 2017
D5	Fire At Esso, 21 August 2017
D6	Vinylchloride Emissie, Shin-Etsu Locatie Botlek-Rotterdam, 23 Augustus 2016
D7	Chemistry in Cooperation - Safety at The Chemelot Industrial Complex
D8	Emission Formaldehyde, 10-22 August 2016
D9	Emission of Ethylene Oxide At Shell Moerdijk
D10	Safety and High-Risk Companies: Lessons Learned After Odfjell
D11	Emissie Van Nikkelstof, BASF Nederland, 6 Februari 2015
D12	Biogasemissie, 7 November 2015
D13	Lekkage Ethyleenoxide
D14	Emission of Fracture Gas (As A Consequence Of Pipe Breach In Heat Exchange), 21...
D15	Two Emissions of Hazardous Substances In 2012
D16	Biogasemissie, 1 April 2014
D17	Explosions Mspo2 Shell Moerdijk
D18	Gas Explosion Diemen
D19	Leakage from Gas Oil Storage Tank, 30 June 2013
D20	Emission of Nickel Dust, 2 August 2012, And 7 October 2013
D21	Earthquake Risks in Groningen
D22	Explosion Ethanol Mixture Vacuum Belt Filter, 11 July 2009
D23	Accident Fatalities in Manure Silo
D24	Fatal Accident Because Of Maintenance of an Oven, Thermphos, Vlissingen, May 15, 2009
D25	Odfjell Terminals Rotterdam Safety, During The 2000 - 2012 Period
D26	Aardgasemissie Bij Gate Terminal Te Rotterdam, 8 September 2011
D27	Explosion Electricity Plant, Nijmegen, 8 November 2012
D28	Iso-Pentaan Emissie, Total Raffinaderij Nederland NV, Nieuwdorp, 1 June 2007
D29	Emissie Uit Salpeterzuurfabriek Bij OCI Nitrogen, Geleen, 9 April 2010

Dutch Safety Board (DSB) Incidents (Continued)

(See: www.onderzoeksraad.nl/en/)

Code	Investigation
D30	Emissie Vloeibaar Ethyleenoxide, Shell Nederland Chemie Bv, Moerdijk, 26 Februari 2009
D31	Emissie Van Propeen In Koelwater, Shell Nederland Chemie BV, Moerdijk, 16 Maart 2007
D32	Fire In Chemical Firm, Moerdijk, 5 January 2011
D33	Fire, Oil Terminal, Bonaire, 8 September 2010
D34	Divinylbenzeen Hp Emissie, Lbc Tank Terminal Bv, Botlek, 27 Oktober 2007
D35	Emissie Stookgas, Shell Nederland Raffinaderij Bv, Pernis, 17 Februari 2007
D36	Koolstofmonoxide-Emissie, Corus Staal BV, Velsen Noord, 15 En 16 Maart 2006
D37	Butaan/Buteen Emissie, Shell Nederland Chemie BV, Pernis, 1 April 2006
D38	Chlooremissie, Akzo Nobel Base Chemicals Bv, 27 Juni 2005
D39	Langdurige Uitstoot Ethyleenoxide, Sterigenics, Zoetermeer
D40	Cast Iron Transport Pipes, Theme Study
D41	Explosion at The Nederlandse Aardoliemaatschappij, Warffum
D42	Labour Accident During Repair of a Gas Leak, Assen
D43	Gas Explosion as a Consequence Of a Failure in the Pipe Coupling, Schijndel
D44	Stroomstoring Haaksbergen, 25 November 2005
D45	Cracked Pipe Causes Dike Collapse, Stein

Health Safety Executive (HSE) of the UK Safety Alerts and Notices—Offshore (English)

(See: www.hse.gov.uk/offshore/notices/sn_index.htm)

Code	Investigation
HA1	Catastrophic failure of shell and tube production cooler
HA2	Flare system impaired by cooling water loss through bursting disc failure on an Intercooler Heat Exchanger
HA3	The maintenance and thorough examination of braking systems on offshore cranes
HA4	Hydrocarbon releases: consideration of acute health effects

Health Safety Executive (HSE) of the UK Safety Alerts and Notices—Offshore (Continued)

(See: www.hse.gov.uk/offshore/notices/sn_index.htm)

Code	Investigation
HA5	'Single line components' in the hoisting and braking systems of offshore cranes
HA6	Weldless repair of safety critical piping systems
HA7	Testing of HVAC dampers (Not Indexable)
HA8	Potential catastrophic failure of pressure-balanced cage-guided control valves and chokes
HA9	Ensuring adequate safety during davit lifeboat drills, testing and maintenance on UK offshore installations (NOT INDEXIBLE)
HA10	Interlocks for drill floor machinery
HA11	Explosion protected electrical heaters

Health Safety Executive (HSE) of the UK Safety Bulletins - COMAH (English)

(See: www.hse.gov.uk/comah/alert.htm)

Code	Investigation
HB1	Explosion and Fire: Chevron Pembroke Refinery, 2 June 2011
HB2	Safety Alert to operators of COMAH oil/fuel storage sites & others storing hazardous substances in large tanks
HB3	Corrosion Fatigue Failure of Tubes in Water Tube Boilers
HB4	Failure of Residual Pressure Valves (RPV) manufactured by Ceodeux Indutec for Transportable Gas Cylinders
HB5	Failure of Residual Pressure Valves (RPV) manufactured by Ceodeux Indutec for Transportable Gas Cylinders - Clarification regarding 14mm plugs PDF
HB6	Explosion in a urea ammonium nitrate (UAN) Fertiliser Transfer Pump
HB7	Safety Alert: Rupture of an (atmospheric) crude oil storage tank PDF
HB8	To operators of COMAH oil/fuel storage sites
HB9	TAV 'checkable' level switches - User safety checks and switch testing

NPO Association for the Study of Failure (ASF) of Japan Incident Database

(For incident reports J1–J163: see www.shippai.org/fkd/en/lisen/cat102.html)

Code	Investigation
J1	Large Explosion at a Fireworks Factory (2003)
J2	Explosion Caused Due to Aged Deterioration of Equipment at an Initiative Explosive Factory (2002)
J3	Fire Due to Accumulated Material in an Exhaust Gas Duct at a Finishing Section of a Synthetic Rubber Plant (2000)
J4	Fire Caused Due to a Flange Loosened from Vibrations at a Synthetic Rubber Plant (2000)
J5	Explosion Due to Inadequate Storage of Explosives in the Temporary Storage House (2000)
J6	Explosion and Fire of Highly Concentrated Hydroxylamine at a Re-Distillation Unit (2000)
J7	Fire of Hexene-1 in the Piping of a Hydrocarbon Vapor Recovery Unit (VRU) at a Tank Lorry Filling Station (2000)
J8	Explosion Caused Due to Unexpected Contaminant During Neutralization Treatment in a Wastewater Tank (1999)
J9	Leakage from a Crack of a Heat Exchanger Due to Corrosion and Abrasion at a Manufacturing Plant of Crude Copper Phthalocyanine Blue (1999)
J10	Leakage and Fire Caused Due to an Abnormal Reaction from Contamination of a Heat Medium to Raw Materials in a Heat Exchanger Type Reactor Having a Corroded Part at an Acrolein Manufacturing Plant (1998)
J11	Explosion Caused Due to Air Contamination During Subdivision Work of Trimethylindium (1998)
J12	Explosion and Fire Caused Due to Mixing of Waste Acids of Different Concentrations in a Waste Acid Tank (1998)
J13	Rupture and Fire of a Measuring Tank Caused Due to an Incompatible Reaction from Transport of a Different Chemical to a Tank Containing a Chemical (1998)
J14	Leakage of Toluene Caused Due to Incorrect Disconnection of a Coupling During Draining of Toluene for Cleaning (1998)
J15	Explosion of Ethanol Vapor Caused Due to Insufficient Exhaust During Drying Operation of Vitamin Tablets at a Pharmaceutical Factory (1998)

NPO Association for the Study of Failure (ASF) of Japan Incident Database (Continued)

(For incident reports J1–J163: see www.shippai.org/fkd/en/lisen/cat102.html)

Code	Investigation
J16	Leakage and Fire of Hydrogen from a Mounting Flange of a Safety Valve in a Reactor at a Succinic Acid Manufacturing Plant (1998)
J17	Explosion of Coke Oven Gas During Cleaning at a Desulfurization Regeneration Tower of a Coke Oven Gas Refining (1998)
J18	Fire of Xylene Remaining in Solid Piperazine Separated in a Centrifuge (1998)
J19	Fire of Ethanol Caused Due to Air Intake in the Ejector of a Treatment Drum at a Surfactant Manufacturing Plant (1998)
J20	Explosion Caused Due to Generation of a Combustible Gas-Air Mixture at a Naphthalene Oxidation Reaction Plant (1998)
J21	Explosion of Acrylic Acid in the Drum Can in the Heating Cabinet for Dissolution (1998)
J22	Damage to a Tank Roof Caused Due to Sticking of a Breather Valve During Transfer of Raw Material (1998)
J23	Explosion of Silicone Products Dissolved in an Organic Solvent During Subdivision Work (1997)
J24	Explosion in the Polycondensation Reaction of Benzyl Chloride (1997)
J25	Ignition of Rubber Remained in the Reactor During Cleaning at a Polybutadiene Manufacturing Plant (1997)
J26	Explosion of a Machine for Melting and Volume Reduction of Polystyrene Foam (1997)
J27	Explosion and Fire Caused Due to Gas Leakage from High-Pressure Ethylene Piping at an Ethanol Manufacturing Plant (1997)
J28	Explosion During Charging Operation of Raw Material Powder into a Reactor Containing Dioxane (1997)
J29	Explosion of an Air Heater of a Boiler at an Agricultural Chemical Manufacturing Plant (1997)
J30	Explosion of an Intermediate Concentration Tank at an Insecticide Manufacturing Plant (1996)
J31	Explosion Due to an Incompatible Reaction in a Nitration Workroom for TNT (1996)
J32	Explosion and Fire Induced Due to Incompatible Reactions of Residual Contaminant at an Alkylaluminium Manufacturing Plant (1996)

NPO Association for the Study of Failure (ASF) of Japan Incident Database (Continued)

(For incident reports J1–J163: see www.shippai.org/fkd/en/lisen/cat102.html)

Code	Investigation
J33	Leakage and Fire Damage Caused Due to Stress Corrosion Cracking of a Heat Exchanger at a Glycerin Concentration Plant (1996)
J34	Explosion Due to Sparks of an Electric Grinder During Repairing a Wastewater Treatment Vessel with Neutralization (1996)
J35	Explosion Caused Due to a Catalytic Effect of Contaminant in the Reactor at a Resin Intermediate Manufacturing Plant (1996)
J36	Fire of a High-Pressure Thermal Gravimeter at a University (1996)
J37	Explosion and Fire at a Manufacturing Plant for Manufacturing a RIM Raw Material Liquid (1996)
J38	Explosion in the Workroom to Fill Explosives with Press (1996)
J39	Fire of Leaked Hydrogen Due to Misuse of a Gasket at a Solvent Manufacturing Plant (1996)
J40	Explosion of Benzene Used as Washing Oil During Cleaning a Fuel Oil Tank (1996)
J41	Explosion in a Storage Tank Caused Due to Decomposition of a Polymerization Catalyst (1995)
J42	Explosion of a Methanol Recovery Residue at an Organic Peroxide Manufacturing Plant (1995)
J43	Explosion and Fire of Tetrahydrofuran During Air-Pressure Transfer to a Tank from a Drum Can (1995)
J44	Ignition of Leaked Gas Caused Due to a Runaway Reaction During a Power Failure at an Expanded Polystyrene Manufacturing Plant (1995)
J45	Explosion Caused Due to a Change of a Draw-Off Method of Coolant in a Jacket of a Reactor from Air Pressure to Steam Pressure at a Medical Intermediate Manufacturing Plant (1995)
J46	Fire Caused Due to Back Flow of Dimethylamine During Repairing Work of Piping from a Reactor to a Scrubber (1995)
J47	Explosion During Taking Out Used Desiccant from a Drum Can (1995)
J48	Rupture, Explosion, and Fire Caused Due to Pressure Rise During Cleaning Operation for a Distillation Kettle (1995)
J49	Explosion During Cleaning by Brushing a Tank for Dangerous Materials at a Paint Manufacturing Factory (1995)

NPO Association for the Study of Failure (ASF) of Japan Incident Database (Continued)

(For incident reports J1–J163: see www.shippai.org/fkd/en/lisen/cat102.html)

Code	Investigation
J50	Fire Caused Due to Overheat of an Electric Heater Caused from Operation Error During a Turnaround Shutdown (1994)
J51	Explosion and Fire at a Butadiene Cylinder Exposed to the Sunshine (1994)
J52	Leakage of Varnish for Ink After Transfer to a Fully Occupied Tank Where an Alarm Lamp Lit (1994)
J53	Explosion and Fire Caused Due to Decomposition of an Organic Peroxide Catalyst (1993)
J54	Fire Caused Due to an Abnormal Reaction at a High Temperature of Waste Oil at the Bottom of a Waste Oil Tank at Styrene Recovery Facilities (1993)
J55	Rupture of a Drum Can Caused by Reactions of Chemicals with Water (1993)
J56	Explosion of a DMSO Recovery Drum Caused Due to Foreign Material Contamination at an Epoxy Resin Manufacturing Plant (1993)
J57	Leakage of Fuel Oil from a Tank and an Outflow to a Canal Caused Due to an Inadequate Automatic Control System (1993)
J58	Leakage of Heat Medium Oil from a Varnish Manufacturing Plant Caused Due to an Error in Judgment and a Mistake in Operation at Startup (1992)
J59	Spouting of Hot Cleaning Liquid from a Manhole Opened for Inspection at an Acrylamide Purifying Column (1992)
J60	Sudden Fire Due to Contamination of Unexpected Impurities During Neutralization Work at a Wastewater Storage Tank (1992)
J61	Explosion at a Fireworks Manufacturing Factory (1992)
J62	Leakage and Explosion of Hydrogen Caused By Using SUS304 in a Chloride Atmosphere and Poor Welding at a Hydrogenated Reduction Plant. (1992)
J63	Explosion Due to a Runaway Reaction from Operation By Misunderstanding of the Reaction Progress During Preparation of Organic Peroxide (1992)
J64	Fire of Methanol Leaked from Piping Removed Accidentally Due to Welding Sparks (1991)

NPO Association for the Study of Failure (ASF) of Japan Incident Database (Continued)

(For incident reports J1–J163: see www.shippai.org/fkd/en/lisen/cat102.html)

Code	Investigation
J65	Explosion of Acetylene Gas Accumulated in a Drum Can of Calcium Carbide On Taking Out (1991)
J66	Eruption Due to a Runaway Reaction from Incorrect Charging Quantity in the Preparation of Acrylic Resin Adhesive (1991)
J67	Fire During Hot Melting Work for a Valve Blocked With Hydrocarbons (1991)
J68	Explosion Caused By Friction On Manufacturing an Air Bag Inflator (1991)
J69	Explosion and Fire Caused By Accumulation of Methyl Hydroperoxide at a Methanol Rectification Column of a Surfactant Manufacturing Plant (1991)
J70	Leakage and Fire of Gas from Lower Piping of a Heating Furnace for Start-Up at an Ammonia Manufacturing Plant (1991)
J71	Explosion and Fire Caused By Insufficient Agitation of Excessive Charging Quantity at a Multi-Purpose Drug Manufacturing Reactor (1991)
J72	Explosion of an Organic Peroxide Catalyst During Circulation Before Use at a Crosslinked Polyethylene Manufacturing Plant (1990)
J73	Fire Occurred Due to Dispersion of Molten Nitrate Caused By High-Pressure Steam Entering a Molten Nitrate Vessel of a Phthalic Anhydride Reactor After Opening a Steam Generation Tube (1990)
J74	Run-Away Reaction Occurred During Vacuum Distillation of Epichlorohydrin Waste Liquid Including Dimethyl Sulfoxide (1990)
J75	Dust Explosion and Fire While Feeding Bisphenol a to a Dissolution Drum (1990)
J76	Explosion in an Intermediate Tank During Turnaround Shutdown Maintenance at a Dimethylformamide Manufacturing Plant (1990)
J77	Rupture of a Reactor Caused By an Abnormal Reaction Due to Lowered Cooling Ability of the Reactor On Manufacturing a Pharmaceutical Intermediate (1990)
J78	Rupture of Metal Drum Cans Containing Extracted Reaction Liquid at a Manufacturing Plant of Phenolic Resin (1990)
J79	Explosion and Fire of Benzoyl Peroxide (BPO) (1990)
J80	Fire in an Electrical Graphitization Furnace for Carbon Fiber Production (1990)

NPO Association for the Study of Failure (ASF) of Japan Incident Database (Continued)

(For incident reports J1–J163: see www.shippai.org/fkd/en/lisen/cat102.html)

Code	Investigation
J81	Explosion of Hydrogen Peroxide Due to the Change of a Feeding Line to a Vessel at a Surfactant Manufacturing Plant (1989)
J82	Explosion and Fire at an Outdoor Tank to Start Storage Before Completion of Attached Facilities (1989)
J83	Explosion of a Dryer Due to Unexpected Reaction of Residual Alkali (1989)
J84	Leakage of Fuel Oil Caused By Damage to a Flexible Hose at a Fuel Oil Tank Piping (1989)
J85	Explosion Caused By an Overflow of Aqueous Hydrogen Peroxide at a Peracetic Acid Manufacturing Plant (1988)
J86	Explosion and Fire D Due to a Change from Sodium Salt to Potassium Salt at a Di-Cumyl Hydroperoxide Manufacturing Plant (1988)
J87	Rupture of a Chlorosulfonic Acid Tank Due to Pressurizing (1988)
J88	Partial Leakage of Hydrochloric Acid Gas from an Absorber Due to an Earthquake (1987)
J89	Fire of Ethylene Oxide Adducts at a Manufacturing Plant Not in Operation (1987)
J90	Rupture of a Solvent Recovery Drum Caused By an Abnormal Reaction Due to a Temperature Rise at a Sugar Ester Manufacturing Plant (1987)
J91	Fire Caused By a Thunderbolt That Struck Piping at a Vinyl Chloride Monomer Manufacturing Plant (1987)
J92	Explosion in Dead Space of a Reactor at a Naphthalene Oxidation Plant (1987)
J93	Explosion of an O-Nitrochlorobenzene Melting Drum Caused Due to a Temperature Rise Caused By Reflux Piping Blockage (1986)
J94	White Fumes Generated from a Toluene Diisocyanate (TDI) Solution Tank Due to Moisture Contamination at an Epichlorohydrin Manufacturer (1986)
J95	Dust Explosion of Purified Anthracene Powder in a Weighing Hopper (1986)
J96	Fire Caused By Electrostatic Charge in the Filtration Process of a Medicine Intermediate (1985)

NPO Association for the Study of Failure (ASF) of Japan Incident Database (Continued)

(For incident reports J1–J163: see www.shippai.org/fkd/en/lisen/cat102.html)

Code	Investigation
J97	Explosion of the Co-Existing System of Epichlorohydrin and Dimethyl Sulfoxide in Epoxy Resin Manufacturing at the Waste Treatment Plant (1985)
J98	Leakage of Toxic Methyl Isocyanate Stored in a Tank at a Chemical Plant (1984)
J99	Explosion of a Germane Gas Cylinder (1984)
J100	Explosion Caused Due to Local Heating of a Drum Can Including Insecticide (Chlorpyrifos-Methyl) (1984)
J101	Explosion at Draw-Off Piping to Remove Hydrocarbons from Liquid Oxygen While Changing Filters at an Air Liquefaction Separator (1984)
J102	Fire of Powder Resin Due to Increased Powder from Remodeling to Improve Efficiency at an ABS Resin Manufacturing Plant (1984)
J103	Explosion and Fire Caused By a Combustible Gas-Air Mixture Generated in a Toluene Drain Tank Sealed With Nitrogen at a Hydroquinone Manufacturing Plant (1984)
J104	Rupture of High-Pressure Air Piping On the Discharge of an Air Compressor (1983)
J105	Hydrogen Chloride Gas Leakage from a Reactor Due to the Pressure Rise from an Abnormal Reaction at a Chloroform Manufacturing Plant (1983)
J106	Warehouse Fire from Explosion of Flammable Gas Released from Expanded Polystyrene Beads Stored in a Warehouse (1982)
J107	Big Explosion Caused By Runaway Reaction of Raw Materials Left After a Small Explosion Due to a Power Failure at an AS Resin Plant (1982)
J108	Explosion of an Acrylic Acid Monomer in a Drum Can During Heating Due to Direct Blowing of Steam (1981)
J109	Fire Due to Mismanagement of Welding at a Warehouse Storing Poisonous Substances Including Sodium Cyanide (1980)
J110	Explosion During Preparation for Manufacturing Pesticide Caused By Stoppage of Cooling Water Due to a Failure Position Setting Error at a Control Valve (1980)
J111	Explosion of 5- Chloro-1,2,3-Thiadiazole (5CT) (1980)

NPO Association for the Study of Failure (ASF) of Japan Incident Database (Continued)

(For incident reports J1–J163: see www.shippai.org/fkd/en/lisen/cat102.html)

Code	Investigation
J112	Explosion in the Piping Caused Due to a Back Flow of Ammonia in the Amination of 5-Chloro-1,2,3-Thiadiazole (1980)
J113	Fire Caused By Explosion of P-Nitrophenol Sodium Salt Due to Friction During Transfer With a Conveyor (1979)
J114	Explosion and Fire Caused By a Runaway Reaction On Start-Up of the Preparation of an Adhesive Manufacturing Plant (1978)
J115	Explosion of DMTP Due to Improper Temperature Control at a Manufacturing Plant of Pesticide Intermediates (1977)
J116	Leakage of Toxic Substances at a Chemical Plant (1976)
J117	Explosion Due to Heat of Adsorption of an Adsorption Tower Used to Deodorize a Methyl Acrylate Tank (1976)
J118	Explosion at a Reactor Caused By a Temperature Rise in a Recovery Process for Hydroxylamine Sulfate at a Pharmaceutical Production Plant (1974)
J119	Disaster of Chemical Plant at Flixborough (1974)
J120	Spouting of High-Temperature Liquid from the Reactor Due to a Hot Spot Formed On Stopping the Agitator (1974)
J121	Rupture of a Vacuum Distillation Drum for 4-Chloro-2-Methylaniline Caused By Air Leakage and Misjudgment (1973)
J122	Leakage and Explosion of a Vinyl Chloride Monomer Due to Valve Damage at Distillation Column Feed Piping at a Plant Manufacturing Vinyl Chloride Monomers (1973)
J123	Fire Caused By Stopping an Agitator in a Liquid Seal-Type Reactor During a Temporary Shutdown Procedure at an Ethylidene Norbornene Plant (1973)
J124	Explosion of 2-Chloropyridine-N-Oxide Left As a Distillation Residue (1973)
J125	Explosion Due to Delayed Start of Agitation at the Start-Up of Reaction of O-Nitrochlorobenzene (1973)
J126	Runaway Reaction During Manufacturing Pesticide Due to a Decomposition Reaction of Tarry Waste (1973)
J127	Explosion Due to Condensation from Miss-Charge of Toluidine Into a Vessel of Diketene (1972)

NPO Association for the Study of Failure (ASF) of Japan Incident Database (Continued)

(For incident reports J1–J163: see www.shippai.org/fkd/en/lisen/cat102.html)

Code	Investigation
J128	Explosion of Dinitrosopentamethylenetetramine (DPT) During Continuous Operation of the Rotary Valve for Measuring DPT Due to Frictional Heat Because of Continuously Abnormal Condition (1972)
J129	Fire Due to Spontaneous Polymerization of Vinyl Acetate During Long-Term Storage (1971)
J130	Explosion Due to Excess Charge of Oxidizer Following Malfunction of Instrument During Manufacturing of Carboxymethylcellulose (1971)
J131	Explosion On Restarting Agitation During the Sulfonation Reaction of Toluene (1970)
J132	Explosion Occurred Immediately After Automation of a Reactor for Preparing Guanidine Nitrate (1970)
J133	Explosion Caused By a Condensation Reaction of Benzyl Chloride Left in a Vessel (1970)
J134	Explosion of 5-T-Butyl-M-Xylene On Restarting an Agitator During the Nitration Reaction (1970)
J135	Spontaneous Ignition of P-Toluenesulfonyl Hydrazone at Temporary Storage (1970)
J136	Explosion of Ozonide Caused By Neglect of a Reaction Due to Improper Temperature Measurement (1969)
J137	Explosion During Transfer of Dioxane By Compressed Air (1969)
J138	Explosion of Acrylic Acid Under Storage in a Drum Can After Partial Melting (1969)
J139	Explosion of the Nitroaniline Preparation Reactor Due to Error in the Quantity of Raw Materials Supplied (1969)
J140	Explosion Caused By Valve Operation Trouble During an Emergency Shutdown of a Compressor at a Vinyl Chloride Plant (1968)
J141	Explosion During the Final Process of Nitration for Nitrocellulose Manufacturing (1968)
J142	Explosion and Fire Due to Improper Raw Material Composition During Preparation of Synthesis of Dehydroacetic Acid (1967)
J143	Explosion in a Cleaning Tank at a Lauroyl Peroxide Manufacturing Plant (1967)

NPO Association for the Study of Failure (ASF) of Japan Incident Database (Continued)

(For incident reports J1–J163: see www.shippai.org/fkd/en/lisen/cat102.html)

Code	Investigation
J144	Explosion of Azobisisobutyronitrile (AIBN) Due to Overheating from Excessive Tightening of Gland Packing of a Drier (1967)
J145	Explosion Caused By Leak of Niter As a Heating Medium in the Distillation Drum at a Phthalic Anhydride Manufacturing Plant (1966)
J146	Self-Ignition of Soft Urethane Foam On Restarting Operation After Maintenance at a Depot (1966)
J147	Burst of a Phenolic Resin Reactor Due to Abnormal Reaction (1965)
J148	Burst of a Reactor Due to Water Leaked Into the P-Nitrotoluene Sulfonic Acid Reactor (1965)
J149	Burst of a Distillation Drum Due to Abnormal Reaction of 2,2'-Dinitrodiphenylamine With Impurities (1964)
J150	Explosion of Organic Peroxide Caused By a Warehouse Fire (1964)
J151	Explosion Due to Hypergolic Hazards in the Intermediate Tank of a Propylene Oxide Manufacturing Plant (1964)
J152	Explosion of Sodium Picramate Due to Slight Stimulation During Handling (1964)
J153	Rupture of the Reactor Due to a Leak of Cooling Water During Sulfonation Reaction of Nitrobenzene (1963)
J154	Explosion Due to Temperature Rise from Careless Handling in the Nitrosation Reaction (1962)
J155	Sudden Explosion of Combustion Gas at a Low-Temperature Liquefaction Separation Plant (1959)
J156	Explosion Caused By a Leakage of a Heat Medium (Nitrate) Through a Crack in the Reactor at a Phthalic Anhydride Manufacturing Plant (1954)
J157	Leakage and Explosion of Benzene Vapor Due to a Pressure Rise in the Reactor Where Steam for Heating Could Not Be Completely Stopped (1956)
J158	Explosion and Fire Caused By Simultaneous Addition of the Whole Quantity of Additives at a Stretch On Redistilling Vinyl Acetate Monomer (1954)
J159	Explosion of Ammonium Nitrate Caused By Superfluous Concentration at an Ammonium Sulfate Recovery Process (1952)

NPO Association for the Study of Failure (ASF) of Japan Incident Database (Continued)

(For incident reports J1–J163: see www.shippai.org/fkd/en/lisen/cat102.html)

Code	Investigation
J160	Explosion Due to Non-Functioning Safety Equipment After an Abnormal Reaction During Manufacturing Phenetidine (1952)
J161	Explosion at a Nitrobenzene Distillation Column Due to the Lowering of the Degree of Vacuum from Power Failure (1951)
J162	Explosion of a Dangerous Material During Repair Work On a Reduction Reactor (1949)
J163	Spontaneous Ignition of Fireworks Including Potassium Chlorate (1943)

(For incident reports J164–J245 see www.shippai.org/fkd/en/lisen/cat103.html)

Code	Investigation
J164	Fire in a Floating Roof Tank of Crude Oil Caused By a Large Earthquake Followed By Fire in Another Floating Roof Tank Two Days Later (2003)
J165	Fire Due to Gasoline Vapor During Simultaneous Remodeling of Two Gasoline Tanks at an Oil Terminal (2003)
J166	Fire While Receiving Gasoline at an Inner Floating Roof Tank (2002)
J167	Leakage and Fire from a Flange With a Special Shape at the Reactor Outlet at a Medium-Pressure Gas Oil Hydrocracker (2002)
J168	Leakage and Fire Caused By Corrosion of Bypass Piping for Recirculation Gas at a Fuel Oil Desulphurization Unit (2002)
J169	Fire Resulting from Hydrogen Sulfide Leak from Overflash Piping of an Atmospheric Distillation Column (2000)
J170	Fire in a Fin-Fan Cooler at A] Reactor Outlet at a Fuel Oil Hydro-Desulfurization Unit (2000)
J171	Fire Due to Cracked Naphtha Leaking from Piping Changed On an Emergency Shutdown (1999)
J172	Gas Oil Leakage from the Tip Valve of a Bottom-Loading Tank Lorry (1999)
J173	Fire at a Desulphurization Plant Due to Leakage of Hydrogen and Gas Oil Mist from a Flange at Outlet Piping of a Reactor Due Caused By a Heavy Rain (1997)

NPO Association for the Study of Failure (ASF) of Japan Incident Database (Continued)

(For incident reports J164–J245 see www.shippai.org/fkd/en/lisen/cat103.html)

Code	Investigation
J174	Fire Caused By an Outflow of Bottom Oil After Opening a Drain Valve at the Bottom of a Column at a Vacuum Distillation Unit (1998)
J175	Fire Due to Leaked Oil from a Heat Exchanger Caused By Slow Action of an Emergency Cut-Off Valve at a Fluidized Catalytic Cracking Unit (1998)
J176	Rupture of a Flush Liquid Cooler for a Mechanical Seal After Repairing a High-Temperature Pump (1997)
J177	Fire of Residue Oil Which Leaked from Carelessly Opened Valves at an Atmospheric Distillation Unit (1997)
J178	Leakage and Fire Caused By Start-Up With a Hose Connected to an Opened Vent Valve at Inlet Piping of a Furnace at a Fuel Oil Desulfurization Unit (1997)
J179	Leakage and Fire Due to Corrosion of Branch Piping for a Thermometer at an Atmospheric Distillation Unit (1996)
J180	Explosion of Hydrogen Gas Due to Backflow During a Turnaround Shutdown at a Hydrogen-Producing Unit. (1996)
J181	Leakage and Fire Due to a Rupture of a Heating Furnace Tube Caused By Unbalanced Heating in a Furnace at a Fuel Oil Direct Desulfurization Unit (1996)
J182	Fire of a Fuel Oil Fraction During Sampling at a Vacuum Distillation Unit (1996)
J183	Fire Caused By Leakage of Heavy Oil from Vibration in the Distillation Section at a Fuel Oil Direct Desulfurization Unit (1996)
J184	Leakage of Crude Oil Due to Abrasion of Crude Oil Piping from External Corrosion (1996)
J185	Leakage of Hydrogen Sulfide Gas Due to Overlapping Errors During Shutdown Maintenance for Off-Site Piping at a Refinery (1995)
J186	Fire Resulting from Leakage of from a Flange Between a Heat Exchanger and Piping During Shutdown Operation at a Vacuum Gas Oil Hydrodesulfurization Unit (1995)
J187	Fire Caused By Flammable Iron Sulfide Generated in a Drainage Tank at an Oil Refinery (1995)

NPO Association for the Study of Failure (ASF) of Japan Incident Database (Continued)

(For incident reports J164–J245 see www.shippai.org/fkd/en/lisen/cat103.html)

Code	Investigation
J188	Fire and Complete Destruction of an Oil Terminal Caused By Using Pipes Before Finishing Repairs (1994)
J189	Leakage of Hydrogen Due to Misuse of a Gasket at a Hydrogenation Desulfurizing Unit of an Oil Refinery (1994)
J190	Leakage of Gasoline Loaded Into a Tank Lorry Due to Disregard of an Over-Loading Alarm (1994)
J191	Explosion and Fire Due to Use of an Electric Spray During Painting Work Inside a Pontoon of a Floating Roof Tank (1994)
J192	Carbon Monoxide Leakage and Fire Due to Breakage of a Gas Expander from Over Revolutions at a Fluidized Catalytic Cracker Unit (1994)
J193	Fire Due to Sudden Rupture of a Buffer Drum On a Compressor at a Light Fuel Oil Desulfurization Unit (1994)
J194	Fuel Oil Leakage from Corroded Branch Piping for a Pressure Gauge Attachment On Jetty Loading Equipment (1993)
J195	Explosion and Fire Caused By the Breakaway of the Cover Plate from the Heat Exchanger of Desulfurization Equipment (1992)
J196	Leakage and Explosion at a Specialized Heat Exchanger Due to Insufficient Maintenance of a Vacuum Gas Oil Hydrodesulfurization Unit (1992)
J197	Fire of Iron Sulfide in an Asphalt Tank During a Turnaround Shutdown (1992)
J198	Fire Occurred After Opening Soil-Covered Underground Military Oil Tank to the Atmosphere (1992)
J199	LPG Leakage and Fire Resulted from Inadequate Repair Work at a Gasification-Desulphurization Plant (1991)
J200	Fire at a Raw Material Feed Pump Caused By Incorrect Operation During Emergency Shutdown Procedure at a Fuel Oil Direct Desulfurization Unit (1991)
J201	Explosion of Residual Gas While Digging Up and Dismantling an Underground Tank at a Gas Station (1991)
J202	Leakage of Fuel Oil Caused By Damage to a Valve Made of Cast Iron at a Refinery (1991)

NPO Association for the Study of Failure (ASF) of Japan Incident Database (Continued)

(For incident reports J164–J245 see www.shippai.org/fkd/en/lisen/cat103.html)

Code	Investigation
J203	Fire Resulted When Iron Sulfide Ignited After Opening a Molten Sulfur Tank (1991)
J204	Oxygen Deficiency Accident Due to Mistaken Connection of Nitrogen Gas Instead of Air During Cleaning of a Kerosene Tank (1980)
J205	Leakage of Water Contaminated With Crude Oil from a Corroded Section of Piping While Removing Unnecessary Piping (1990)
J206	Fire Due to Heat Accumulation from Oxidation of Materials Stuck On the Wall of a Blown Asphalt Tank at a Refinery (1990)
J207	Leakage of Fuel Oil Into the Sea from a Corroded Section of Receiving Piping Under Hot Insulation at a Jetty (1990)
J208	Leakage Caused By Delay in Valve Operation While Loading Kerosene Into a Tanker (1990)
J209	Naphtha and Hydrogen Leakage and Fire Caused By a Problem Tightening a Flange at a Naphtha Catalytic Reforming Unit (1990)
J210	Fire Resulted After Vacuum Residue Leaked from a Pipe Due to Corrosion in the Bottom Recycling Line at a Vacuum Distillation Unit. (1990)
J211	Leakage and Fire of Heavy Residue Oil from a Drain Valve After Pump Switching at a Vacuum Distillation Unit of a Refinery (1990)
J212	Gas Oil Fire Caused By Leakage from a Bonnet Flange of a Valve Usually Not Used at an Atmospheric Distillation Unit (1990)
J213	Fire Caused By Gas Leaking from a Piping Flange at a Vacuum Gas Oil Hydro-Desulfurization Unit (1989)
J214	Leakage and Explosion of Hydrogen at Outlet Piping of a Reactor in the Indirect Hydrodesulfurization Unit of a Fuel Oil (1989)
J215	Fire Caused Due to Erosion of a Water Injection Nozzle to Reactor Outlet Piping at a Heavy Oil Hydrodesulfurization Unit (1988)
J216	Fire and Explosion of High-Pressure Hydrogen Caused Due to Failure of a Stopping Operation of a Feed Pump of Raw Oil at a Demonstration Unit for Catalytic Cracking of a Residue (1988)
J217	Sinking of a Floating Roof Due to Inundating of Pontoons and Retained Rainwater On the Roof at a Floating Roof Naphtha Tank (1987)

NPO Association for the Study of Failure (ASF) of Japan Incident Database (Continued)

(For incident reports J164–J245 see www.shippai.org/fkd/en/lisen/cat103.html)

Code	Investigation
J218	Fire Caused By Leakage from a Control Valve Flange, Which Became Loose Due to Vibration at a Catalytic Reforming Unit (1987)
J219	Fire at a Fin-Fan Cooler Due to Contamination of Hydrogen By Mishandling of a Valve During Turnaround Shutdown Maintenance at an Atmospheric Distillation Unit (1987)
J220	Gasoline Fire During Drain Work at a Filter of Lorry Shipping Facilities (1987)
J221	Explosion at a Furnace Caused By Fuel Gas Leakage On Start-Up of a Catalytic Reforming Unit (1986)
J222	Explosion and Fire at a Chemical Tanker During Cargo Handling Due to Static Electricity from an Insulated Level Gage (1985)
J223	Fire Due to Residual Oil During Repairs at a Fuel Oil Tank of Refinery (1984)
J224	Leakage Due to a Cracked Weld On a Block Valve for Branched Piping at a Fuel Oil Hydrodesulfurization Unit (1984)
J225	Explosion and Fire in a Distillation Tank at a Waste Oil Regeneration Factory (1983)
J226	Leakage of Gas Oil from a Crack in the Corroded Part of a Tank Side Wall Due to an Earthquake (1983)
J227	Burst By Hydrogen Attack On Residue in Hydrodesulfurization Unit Piping (1982)
J228	Explosion While Sampling of Kerosene Fractions from a Tank Roof (1981)
J229	Fire in a Soil-Covered Underground U.S. Army Oil Tank Caused By Hot Work at an Adjacent Tank (1981)
J230	Brittle Fracture of Hydrodesulfurization Reactor During Pressure Test (1980)
J231	Rupture During an Air-Tight Test of Reactor Used for 21 Years at a Catalytic Hydro-Desulfurization Unit (1980)
J232	Leakage and Fire of Heavy Gas Oil from an Opening in Reflux Pump Vent Piping at a Distillation Column (1978)
J233	Outflow of All Fuel Oil from a Tank Caused By Breakage of a Base Plate of an Outdoor Oil Tank from an Earthquake (1978)

NPO Association for the Study of Failure (ASF) of Japan Incident Database (Continued)

(For incident reports J164–J245 see www.shippai.org/fkd/en/lisen/cat103.html)

Code	Investigation
J234	Explosion and Fire of a Power Generation Boiler On Restarting Operation (1977)
J235	Leakage and Fire of Hydrogen Due to Stress Corrosion Cracking That Originated from the Influence of Turnaround Shutdown Maintenance On a Drain Valve at Hydrogen Gas Piping in a Fuel Oil Desulfurization Cracking Unit (1999)
J236	Explosion of a Toluene Tank Due to Static Electricity On Sampling (1976)
J237	Explosion and Fire of Light Oil Evaporated at a Waste Oil Regeneration Factory (1975)
J238	Fire Due to Structure Dropped Into a Corroded Kerosene Tank (1975)
J239	Oil Spill By Mizushima's Tank Damage (1974)
J240	Outflow of 43000 Kl Fuel Oil Into Setonaikai Sea Due to a Crack at the Base Plate of a Tank (1974)
J241	Fire Due to Damaged Glass in a Level Gauge at a Hydrogen Sulfide Absorption Column at a Kerosene Hydrodesulfurization Unit (1973)
J242	Explosion and Fire On Sampling at a Benzene Tank (1972)
J243	Leakage of a Newly Constructed Floating Roof Tank Due to Damage to a Base Plate During a Water Leak Test (1968)
J244	Explosion and Fire of LPG Tanks (1966)
J245	Fire of Petroleum Tank, Etc. By Niigata Earthquake (1964)

(For incident reports J246–J271 see www.Shippai.Org/Fkd/En/Lisen/Cat103.html)

Code	Investigation
J246	Turf Grass Fire Caused By Scatter from a Flare Stack (2000)
J247	Fire During Drain Work Caused By Generating an Insulated State On Hanging a Bucket On the Valve of an Intermediate Raw Material Drum for Hydrocarbon Resin (2000)
J248	Leakage of High-Pressure Ethylene Gas from Gland Packing of an Acetone Injection Valve at a Low-Density Polyethylene Manufacturing Plant (1999)

NPO Association for the Study of Failure (ASF) of Japan Incident Database (Continued)

(For incident reports J246–J271 see www.Shippai.Org/Fkd/En/Lisen/Cat103.html)

Code	Investigation
J249	Leakage of Agitator Hydraulic Oil Caused By Increase in Reaction Fluid Viscosity from Contamination of a Catalyst in a Polystyrene Polymerization Reactor (1998)
J250	Leakage and Fire of Gas from a High-Temperature Heat Exchanger at a Benzene Manufacturing Plant Using a Dealkylation Reaction (1998)
J251	Deformation of a Kerosene Tank in Which a Blind Plate Was Inserted By Mistake (1996)
J252	Fire Caused By Incorrect Opening of Valves During Sampling Operation at a Polypropylene Manufacturing Plant (1996)
J253	Overflow of Gasoline from a Plain Gasoline Tank at a Factory (1995)
J254	Ink Varnish Leakage from Trench Piping at a Lorry Station (1994)
J255	Leakage and Fire of Hydrogen During Exchange of a Dehydrogenation Catalyst at an Alkylbenzene Manufacturing Plant (1994)
J256	Leakage and Fire from a Reactor Flange During Shutdown Operation at a Hydrogenation Plant of Pyrolysis Gasoline (1993)
J257	Fire of Naphtha Caused By an Operator Who Went Another Site to Do Other Work Leaving a Drain Valve Open at an Ethylene Plant (1992)
J258	Fire of Ethylene Leaked from Mounting Part of a Rupture Disk of a Polymerization Reactor at a High-Pressure Polyethylene Plant (1991)
J259	Explosion Caused By a High Temperature from Adiabatic Compression While Transferring High-Pressure Air at a Polyethylene Manufacturing Plant (1991)
J260	Fire During Renewal Work On Packing Materials of a Gasoline Column at an Ethylene Manufacturing Plant (1991)
J261	Fire of Heptane Due to Improper Valve Handling at a Polypropylene Manufacturing Plant (1989)
J262	Leakage of Raw Material Oil from a Hydrocarbon Resin Reactor at the Beginning of a Reaction (1989)
J263	Leakage and Fire of Naphtha Caused Due to Overlooking of an Open Drain Valve at an Ethylene Plant (1989)

NPO Association for the Study of Failure (ASF) of Japan Incident Database (Continued)

(For incident reports J246–J271 see www.Shippai.Org/Fkd/En/Lisen/Cat103.html)

Code	Investigation
J264	Fire and Explosion in a Butadiene Rectifying Column During Preparation for Turnaround Shutdown Maintenance (1988)
J265	Small Leakage of Naphtha from an Opening Caused By Corrosion of Air Piping for Decoking at an Ethylene Manufacturing Unit (1986)
J266	Fire of Cracked Gas Leaked from Outlet Piping of an Ethane Cracking Furnace at an Ethylene Manufacturing Plant (1985)
J267	Explosion Due to Flame During Welding to Install Piping to a Neutralization Tank at a Hydrocarbon Resin Plant (1983)
J268	Fire from a Vent Stack of an Extruder During a Shutdown at a Polyethylene Manufacturing Plant (1973)
J269	Fire Due to Inadequate Finishing of an O-Ring Seal at a Polyethylene Manufacturing Plant Using a High-Pressure Method (1973)
J270	Explosion Caused By Mishandling of Only One Remote Control Valve Separated from Atmosphere in the Polymerization Reactor at a Polypropylene Manufacturing Plant (1973)
J271	Fire at an Acetylene Hydrogenation Section On Rapid Re-Startup After an Emergency Shutdown at an Ethylene Plant (1973)

Standalone Special Inquiry Reports

	Investigation	Source
S1	Piper Alpha	www.hse.gov.uk/offshore/piper-alpha-disaster-public-inquiry.htm
S2	Jaipur	Multiple links starting with: www.oisd.gov.in/Image/GetDocumentAttachmentByID?documentID=73 (Then use the same link, but ending in 74–87, to get all chapters/sections
S3	Buncefield	www.hse.gov.uk/comah/buncefield/buncefield-report.pdf
S4	Seveso	www.hse.gov.uk/comah/sragtech/caseseveso76.htm

Standalone Special Inquiry Reports

	Investigation	Source
S5	Montara oil spill	www.iadc.org/wp-content/uploads/2016/02/201011-Montara-Report.pdf
S6	PDVSA Amaguay	Report has been redacted – not currently available
S7	Hickson Welch	www.hse.gov.uk/comah/sragtech/casehickwel92.htm
S8	Apollo 1 Fire	history.nasa.gov/Apollo204/appendices/AppendixD12-17.pdf
S9	Challenger Shuttle	spaceflight.nasa.gov/outreach/SignificantIncidents/assets/rogers_commission_report.pdf
S10	Columbia Shuttle	www.nasa.gov/columbia/home/CAIB_Vol1.html
S11	Lodi	archive.epa.gov/emergencies/docs/chem/web/pdf/napp.pdf
S12	Bhopal	www.csb.gov/on-30th-anniversary-of-fatal-chemical-release-that-killed-thousands-in-bhopal-india-csb-safety-message-warns-it-could-happen-again-/
S13	Bass Strait	www.nopsema.gov.au/assets/News-and-media/Announcement-Investigation-into-fatalities-on-Stena-Clyde-drilling-rig-Bass-Strait-17-October-2012.pdf
S14	Port Neal	archive.epa.gov/emergencies/docs/chem/web/pdf/cterra.pdf
S15	Lac Megantic	www.tsb.gc.ca/eng/rapports-reports/rail/2013/R13D0054/R13D0054.pdf
S16	Brumadinho	www.anm.gov.br/parecer-007-2019-brumadinho-final and www.b1technicalinvestigation.com/
S17	Firestone	ntsb.gov/investigations/AccidentReports/Reports/PAB1902.pdf

A.5 References

A.1 API Recommended Practice (RP) (2016). *Process Safety Performance Indicators for the Refining and Petrochemical Industries.* Washington, DC: American Petroleum Institute.

A.2 CCPS (2009). *Guidelines for Process Safety Metrics*. Hoboken, NJ: AIChE/Wiley.

A.3 CCPS (2011). *Guidelines for Auditing Process Safety Management Systems*, 2nd Edition. Hoboken, NJ: AIChE/Wiley.

A.4 CCPS (2018). *Process Safety Metrics: Guide for Selecting Leading and Lagging Indicators*, Version 3.2. New York: AIChE.

Index